A PROJECT MANAGER'S
BOOK OF FORMS

A PROJECT MANAGER'S BOOK OF FORMS

A Companion to the *PMBOK*® *Guide*— Fourth Edition

Cynthia Snyder Stackpole

WILEY

John Wiley & Sons, Inc.

Project Management Institute

Library of Congress Cataloging-in-Publication Data:

Snyder Stackpole, Cynthia, 1962-
    A project manager's book of forms : a companion to the PMBOK guide—fourth edition / Cynthia Snyder Stackpole.
      p. cm.
    Includes index.
    ISBN 978-0-470-38984-3 (paper/cd)
      1. Project management—Forms.   I. Guide to the project management body of knowledge (PMBOK guide).
    II. Title.
  HD69.P75S689   2009
  658.4'04—dc22

2008044563

Printed in the United States of America

10 9 8 7 6 5 4

Contents

Acknowledgments

To Ruth Anne Guerrero: Thank you for your keen insight and your support for my work on the *PMBOK® Guide—Fourth Edition* and this companion book of forms. Your energy and laser-sharp thinking have contributed vastly. It has been such a pleasure to work with you.

To Joseph Kestel: I appreciate your time in reviewing the proposal for this book and providing thoughtful input to the approach. I also appreciate your wonderful leadership skills on Chapters 3 and 5 of the *PMBOK® Guide—Fourth Edition*.

To Dan Picard: You are a rare find, an editor and a PMP! Your insight into both worlds is so valuable. I really appreciate your taking the time to give me your unique perspective for this book and the others we have worked on. Long live AJ!

To Terri Gaydon: How do I say thank you to a friend, a mentor, and an advisor? Your questions, insights, and perspectives were so valuable in putting this book together. And, of course, over the past 15 years your friendship has meant the world to me. Thank you a million times.

A PROJECT MANAGER'S BOOK OF FORMS

Introduction

The *Project Management Book of Forms* is designed to be a companion to *A Guide to the Project Management Body of Knowledge (PMBOK® Guide)*—Fourth Edition. The purpose is to present the information from the *PMBOK® Guide*—Fourth Edition in a set of forms and reports so that project managers can readily apply the concepts and practices described in the *PMBOK® Guide*—Fourth Edition to their projects.

The *PMBOK® Guide*—Fourth Edition identifies that subset of the project management body of knowledge generally recognized as good practice. As an ANSI Standard, it does not describe how to apply those practices, nor does it provide a vehicle for transferring that knowledge into practice.

This *Book of Forms* will assist project managers in applying information presented in the *PMBOK® Guide*—Fourth Edition into project documentation. The *Book of Forms* does not teach project management concepts or describe how to apply project management techniques. Textbooks and classes can fulfill those needs. This book provides an easy way to apply good practices to projects.

Since one of the defining factors about projects is that they are unique, project managers must tailor the forms and reports to meet the needs of their individual projects. Some projects will require information in addition to what is presented in these forms; some will require less. These forms are presented in paper format and electronic versions to make them easy to adapt to the needs of specific projects. They follow the information in the *PMBOK® Guide*—Fourth Edition but can be adapted to meet the needs of the project manager and specific projects.

AUDIENCE

This book is written specifically for project managers to help manage all aspects of the project. Those new to project management can use the forms as a guide in collecting and organizing project information. Experienced project managers can use the forms as a template so that they collect a set of consistent data on all projects. In essence, the forms save reinventing the wheel for each project.

A secondary audience is the manager of project managers or a project management office. Using the information in this book ensures a consistent approach to project documentation. Adopting these forms on an organizational level will enable a repeatable approach to project management.

ORGANIZATION

The forms are organized by process group: initiating, planning, executing, monitoring and controlling, and closing. Within those process groups, the forms are arranged sequentially as presented in the *PMBOK® Guide*—Fourth Edition.

A description of each form is presented along with a list of contents. For the planning forms, there is a description of where the information in the form comes from (inputs) and where it goes to (outputs). For some forms, there is a list of related forms. On the page(s) to come, a blank copy of the form is presented, followed by a copy of the form with a description of the information that goes into each field. In the back of the book, there is a completely editable CD-ROM with a copy of all the blank forms. All forms are in Microsoft® Office software for ease of tailoring.

Some forms are included that are not mentioned in the *PMBOK® Guide*—Fourth Edition. These are forms that assist in managing a project but are not considered part of the project management standard.

Not all forms will be needed on all projects. Use the forms you need, to the degree that you need them, to assist you in managing your projects.

2

Initiating Forms

2.1 INITIATING PROCESS GROUP

The purpose of the Initiating Process Group is to authorize a project, provide a high-level definition of the project, and identify stakeholders. There are two processes in the Initiating Process Group:

- Develop Project Charter
- Identify Stakeholders

The intent of the Initiating Process Group is to at least:

- Authorize a project
- Identify project objectives
- Define the initial scope of the project
- Obtain organizational commitment
- Assign a project manager
- Identify project stakeholders

As the first processes in the project, the initiating processes are vital to starting a project effectively. These processes can be revisited throughout the project for validation and elaboration as needed.

The forms used to document initiating information include:

- Project Charter
- Stakeholder Register
- Stakeholder Analysis Matrix
- Stakeholder Management Strategy

These forms are consistent with the information in the *PMBOK® Guide*—Fourth Edition. Tailor them to meet the needs of your project by editing, combining, or revising them.

2.2 PROJECT CHARTER

The Project Charter is a document that formally authorizes a project or phase. The Project Charter defines the reason for the project and assigns a project manager and his or her authority level for the project. The contents of the charter describe the project in high-level terms, such as:

- Purpose or justification
- High-level project description
- High-level project and product requirements
- Summary budget
- Summary milestone schedule
- Initial risks
- Project objectives and success criteria
- Acceptance criteria
- Project manager authority

Use the information from your project to tailor the form to best meet your needs.

The Project Charter can receive information from:

- Contracts
- Statements of work
- Business case

It provides information to:

- Project Management Plan
- Project Scope Statement
- Stakeholder Register
- Requirements Documentation
- Requirements Management Plan
- Requirements Traceability Matrix

The Project Charter is an output from the process 4.1 Develop Project Charter in the *PMBOK® Guide*—Fourth Edition.

PROJECT CHARTER

Project Title: _____

Project Sponsor: _____ Date Prepared: _____

Project Manager: _____ Project Customer: _____

Project Purpose or Justification:

Project Description:

High-level Project and Product Requirements:

Summary Budget:

Initial Risks:

PROJECT CHARTER

Summary Milestones	Due Date

Project Objectives	Success Criteria	Person Approving
Scope:		
Time:		
Cost:		
Quality:		
Other:		

PROJECT CHARTER

Acceptance Criteria:

Project Manager Authority Level
Staffing Decisions:

Budget Management and Variance:

Technical Decisions:

Conflict Resolution:

Escalation Path for Authority Limitations:

Approvals:

Project Manager Signature	Sponsor or Originator Signature
Project Manager Name	Sponsor or Originator Name
Date	Date

PROJECT CHARTER

Project Title: _____

Project Sponsor: _____ **Date Prepared:** _____

Project Manager: _____ **Project Customer:** _____

Project Purpose or Justification:

Define the reason the project is being undertaken. This section may refer to a business case, the organization's strategic plan, external factors, a contract or any other document or reason for performing the project.

Project Description:

Provide a summary-level description of the project. This section may include information on high-level product and project deliverables as well as the approach to the project.

High-level Project and Product Requirements:

Define the high-level conditions or capabilities that must be met to satisfy the purpose of the project. Describe the product features and functions that must be present to meet stakeholders' needs and expectations. This section does not describe the detailed requirements as those are covered in requirements documentation.

Summary Budget:

List the initial range of estimate expenditures for the project.

Initial Risks:

Document initial project risks. These will later be entered into a Risk Register when project planning begins.

PROJECT CHARTER

Summary Milestones	Due Date
List the significant events in the project. These can include the completion of key deliverables, the beginning or completion of a project phase or product acceptance.	*Completion date of the milestone.*

Project Objectives	Success Criteria	Person Approving

Scope:

A statement that describes the scope needed to achieve the planned benefits of the project.	*The specific and measureable criteria that will determine project scope success.*	*The name or position of the person that can sign off on the scope objectives.*

Time:

A statement that describes the goals for the timely completion of the project.	*The specific dates that must be met to determine schedule success.*	*The name or position of the person that can sign off on the schedule objectives.*

Cost:

A statement that describes the goals for the project expenditures.	*The specific currency or range of currency that defines budgetary success.*	*The name or position of the person that can sign off on the cost objectives.*

Quality:

A statement that describes the quality criteria for the project.	*The specific measurements that must be met for the project and product to be considered a success.*	*The name or position of the person that can sign off on the quality objectives.*

Other:

Any other types of objectives appropriate to the project.	*Relevant specific measureable results that define success.*	*The name or position of the person that can sign off on the objectives.*

PROJECT CHARTER

Acceptance Criteria:

Identify the criteria that must be met in order for the project to be accepted by the customer or sponsor.

Project Manager Authority Level
Staffing Decisions:

Define the authority of the project manager to hire, fire, discipline, accept or not accept project staff.

Budget Management and Variance:

Define the authority of the project manager to commit, manage, and control project funds. Include variance levels that require escalation for approval or re-baselining.

Technical Decisions:

Define the authority of the project manager to make technical decisions about the deliverables or the project approach.

Conflict Resolution:

Define the authority of the project manager to resolve conflict within the team, within the organization, and with external stakeholders.

Escalation Path for Authority Limitations:

Define the path of escalation for issues outside the authority level of the project manager.

Approvals:

_____
Project Manager Signature

_____
Project Manager Name

_____
Date

_____
Sponsor or Originator Signature

_____
Sponsor or Originator Name

_____
Date

2.3 STAKEHOLDER REGISTER

The Stakeholder Register is used to identify those people and organizations impacted by the project and document relevant information about each stakeholder. Relevant information can include:

- Name
- Position in the organization
- Role in the project
- Contact information
- List of stakeholder's major requirements
- List of stakeholder's expectations
- Potential influence on the project
- A classification or categorization of each stakeholder

Information in the Stakeholder Register should be tailored to meet the needs of the project. For example, some projects may have internal and external stakeholders while others may only have internal stakeholders. Some projects may categorize stakeholders as friend, foe, or neutral; others may categorize them as high, medium, or low influence. The sample on the next page is just one approach to identifying and documenting stakeholder information.

Use the information on your project to tailor the form to best meet your needs.

The Stakeholder Register receives information from:

- Project Charter
- Procurement documents

It is related to:

- Stakeholder Analysis Matrix
- Stakeholder Management Strategy

It provides information to:

- Requirements Documentation
- Quality Management Plan
- Risk Register

The Stakeholder Register is an output from the process 10.1 Identify Stakeholders in the *PMBOK® Guide—Fourth Edition.*

STAKEHOLDER REGISTER

Project Title: _____

Date Prepared: _____

Name	Position	Role	Contact Information	Requirements	Expectations	Influence	Classification

STAKEHOLDER REGISTER

Project Title: _____

Date Prepared: _____

Name	Position	Role	Contact Information	Requirements	Expectations	Influence	Classification
Stakeholder's name.	*Position in the organization.*	*The function they perform on the project.*	*Communication and correspondence information.*	*High-level needs or wants for the project and/or product.*	*Expectations of the project or product.*	*Level and type of influence on the project.*	*A category or classification.*

2.4 STAKEHOLDER ANALYSIS MATRIX

The Stakeholder Analysis Matrix is used to categorize stakeholders. It can be used to help fill in the Stakeholder Register. The categories of stakeholders can also assist in developing stakeholder management strategies that can be used for groups of stakeholders.

The example on the next page is used to assess the relative power (high or low) on one axis and the relative interest (high or low) on the other axis. There are many other ways to categorize stakeholders using a grid. Some examples include:

- Influence/impact
- Friend/foe

The needs of the project will determine if a Stakeholder Analysis Matrix will be helpful and, if so, what stakeholder aspects should be assessed.

Use the information from your project to tailor the form to best meet your needs.

The Stakeholder Analysis Matrix receives information from:

- Project Charter
- Procurement documents

It is related to:

- Stakeholder Register
- Stakeholder Management Strategy

It provides information to:

- Communications Management Plan

The Stakeholder Analysis Matrix is an output from the process 10.1 Identify Stakeholders in the *PMBOK® Guide*—Fourth Edition.

STAKEHOLDER ANALYSIS MATRIX

Project Title: _____   Date Prepared: _____

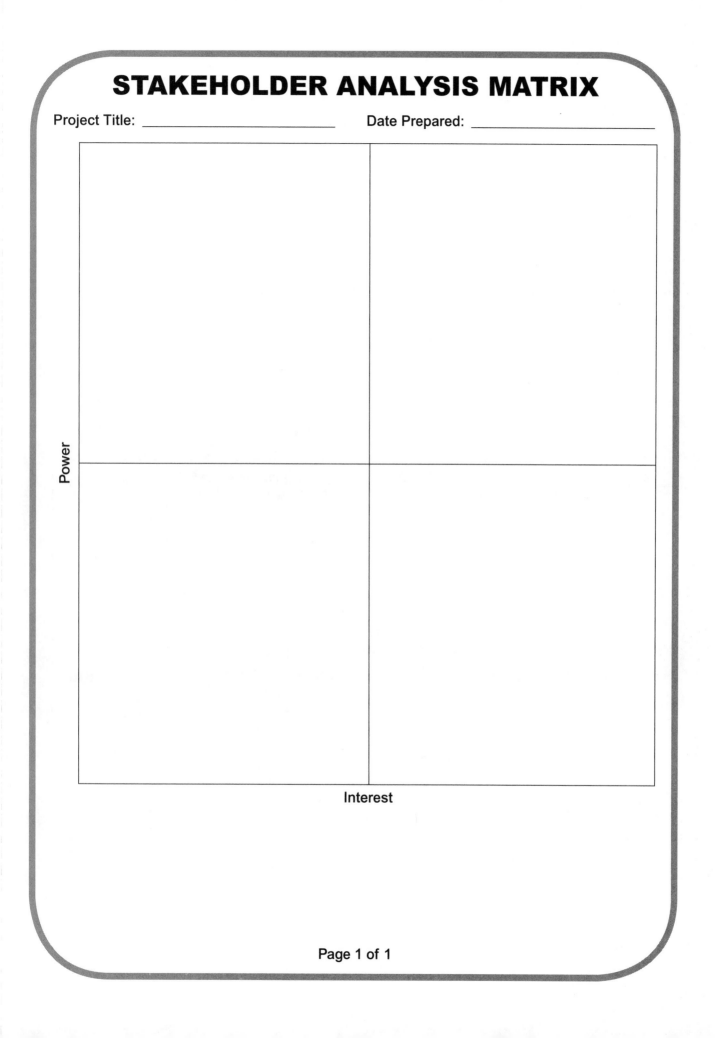

Power

Interest

STAKEHOLDER ANALYSIS MATRIX

Project Title: _____ Date Prepared: _____

Power

Place stakeholders with high power and low interest in the project here.	*Place stakeholders with high power and high interest in the project here.*
Place stakeholders with low power and low interest in the project here.	*Place stakeholders with low power and high interest in the project here.*

Interest

2.5 STAKEHOLDER MANAGEMENT STRATEGY

Stakeholder Management Strategy documents stakeholders and their influence on the project and analyzes the impact that they can have on the project. It also provides a place to document potential strategies to increase stakeholders' positive influence and minimize potential disruptive influence on the project.

This type of document may not be needed on all projects. On some projects, it may be combined with the Stakeholder Register. Information in this document may be considered sensitive. Therefore, the project manager should consider how much information to document and how widely to share the information.

Use the information from your project to tailor the form to best meet your needs.

Stakeholder Management Strategy receives information from:

* Project Charter
* Procurement documents

It is related to:

* Stakeholder Register
* Stakeholder Analysis Matrix

It provides information to:

* Communications Management Plan

The Stakeholder Management Strategy is an output from the process 10.1 Identify Stakeholders in the *PMBOK® Guide*—Fourth Edition.

STAKEHOLDER MANAGEMENT STRATEGY

Project Title: _____

Date Prepared: _____

Name	Influence	Impact Assessment	Strategies

STAKEHOLDER MANAGEMENT STRATEGY

Project Title: _____

Date Prepared: _____

Name	Influence	Impact Assessment	Strategies
Name of stakeholder.	*Type of influence.*	*Degree of influence or impact of influence.*	*Strategies and tactics to maximize positive stakeholder influence and minimize or neutralize negative stakeholder influence.*

Planning Forms

3.1 PLANNING PROCESS GROUP

The purpose of the Planning Process Group is to elaborate the information from the Project Charter to create a comprehensive set of plans that will enable the project team to deliver the project objectives. There are 20 processes in the Planning Process Group.

- Develop Project Management Plan
- Collect Requirements
- Define Scope
- Create WBS
- Define Activities
- Sequence Activities
- Estimate Activity Resources
- Estimate Activity Durations
- Develop Schedule
- Estimate Costs

- Determine Budget
- Plan Quality
- Develop Human Resource Plan
- Plan Communications
- Plan Risk Management
- Identify Risks
- Perform Qualitative Analysis
- Perform Quantitative Analysis
- Plan Risk Responses
- Plan Procurements

The intent of the Planning Process Group is to at least:

- Elaborate and clarify the project scope
- Develop a realistic schedule
- Develop a realistic budget
- Identify project and product quality processes
- Plan the human resource aspects of the project
- Determine the communication needs
- Establish risk management practices
- Identify the procurement needs of the project
- Combine all the planning information into a Project Management Plan and a set of project documents that are cohesive and integrated

Planning is not a one-time event. It occurs throughout the project. Initial plans will become more detailed as additional information about the project becomes available. Additionally, as changes are approved for the project or product, many of the planning processes will need to be revisited and the documents revised and updated.

Many of the forms in this section provide information needed for other forms. The form description indicates where information is received from and where it goes to.

The forms used to document planning information include:

- Requirements Documentation
- Requirements Management Plan
- Requirements Traceability Matrix
- Project Scope Statement
- Assumption and Constraint Log
- Work Breakdown Structure
- Work Breakdown Structure Dictionary
- Activity List
- Activity Attributes
- Milestone List
- Network Diagram
- Activity Resource Requirements
- Resource Breakdown Structure
- Activity Duration Estimates
- Duration Estimating Worksheet
- Project Schedule
- Activity Cost Estimates
- Cost Estimating Worksheet
- Bottom-up Cost Estimating Worksheet
- Cost Performance Baseline
- Quality Management Plan
- Quality Metrics
- Process Improvement Plan
- Responsibility Assignment Matrix
- Roles and Responsibilities
- Human Resource Plan
- Communications Management Plan
- Risk Management Plan
- Risk Register
- Probability and Impact Assessment
- Probability and Impact Matrix
- Risk Data Sheet
- Procurement Management Plan
- Source Selection Criteria
- Project Management Plan
- Configuration Management Plan
- Change Management Plan

Some forms in this section are not explicitly described in the *PMBOK® Guide*—Fourth Edition, but they are useful in planning and managing a project. The forms described in *PMBOK® Guide* are consistent with the Fourth Edition. Tailor all forms to meet the needs of your project by editing, combining, or revising them.

3.2 REQUIREMENTS DOCUMENTATION

Project and product requirements need to be documented. In addition to documenting requirements, it is useful to document the stakeholder associated with the requirements, categorize and prioritize requirements, and define the acceptance criteria. This documentation assists the project manager in making trade-off decisions among requirements and in managing stakeholder expectations. Requirements will be progressively elaborated as more information about the project becomes available. Other information about requirements may be documented, such as:

* Relationship between requirements
* Impacts of requirements
* Assumptions and constraints

When documenting requirements, it useful to group them by category. Some common categories include:

* Functional requirements
* Quality requirements
* Performance requirements
* Safety requirements
* Security requirements
* Technical requirements
* Training requirements
* Support and maintainability requirements

Use the information from your project to tailor the form to best meet your needs.

Requirements documentation can receive information from:

* Project Charter
* Stakeholder Register

It is related to:

* Requirements Management Plan
* Requirements Traceability Matrix

It provides information to:

* Project Scope Statement
* Work Breakdown Structure
* Project Management Plan
* Accepted deliverables
* Change Requests

Requirements documentation is an output from the process 5.1 Collect Requirements in the *PMBOK®* *Guide*—Fourth Edition.

REQUIREMENTS DOCUMENTATION

Project Title: _____   Date Prepared: _____

Stakeholder	Requirement	Category	Priority	Acceptance Criteria

REQUIREMENTS DOCUMENTATION

Project Title: _____   Date Prepared: _____

Stakeholder	Requirement	Category	Priority	Acceptance Criteria
Identify the name or organization of the stakeholder.	*Identify the requirement.*	*Assign a category.*	*Prioritize in total or by category.*	*Define the criteria for acceptance.*

3.3 REQUIREMENTS MANAGEMENT PLAN

The Requirements Management Plan is part of the Project Management Plan. It specifies the way that requirements will be managed throughout the project. Managing requirements includes at least:

- Collecting
- Categorizing
- Prioritizing
- Managing change
- Verifying

It can also include how they will be traced and related on a Requirements Traceability Matrix and the method used to validate that they meet stakeholders' expectations.

Use the information from your project to tailor the form to best meet your needs.

The Requirements Management Plan can receive information from:

- Project Charter
- Stakeholder Register

It is related to:

- Requirements Documentation
- Requirements Traceability Matrix

It provides information to:

- Project Management Plan

The Requirements Management Plan is an output from the process 5.1 Collect Requirements in the *PMBOK® Guide*—Fourth Edition.

REQUIREMENTS MANAGEMENT PLAN

Project Title: _____   Date: _____

Requirements Collection:

Categories:

Prioritization:

Traceability:

Configuration Management:

Verification:

REQUIREMENTS MANAGEMENT PLAN

Project Title: _____ Date: _____

Requirements Collection:

Describe how requirements will be collected. Consider such techniques as brainstorming, interviewing, observation, etc.

Categories:

Identify the categories that will be used to group requirements.

Prioritization:

Identify the approach to prioritize requirements.

Traceability:

Identify the requirement attributes that will be used for tracing requirements, such as functional to business requirements or functional to security requirements.

Configuration Management:

Describe how requirements can be changed. Include a description of the process and any necessary forms, processes, or procedures needed to initiate a change. Document how analysis of the impact of changes will be conducted. Include levels of approval necessary for changes.

Verification:

Describe the different methods that will be used to verify requirements, such as observation, measurement, testing, etc. Include any metrics that will be used for verification.

3.4 REQUIREMENTS TRACEABILITY MATRIX

A Requirements Traceability Matrix is used to track the various attributes of requirements throughout the project life cycle. It uses information from Requirements Documentation and traces how those requirements are addressed through other aspects of the project. The form on the next page shows how requirements would be traced to project objectives and deliverables and how they will be verified and validated.

Another way to use the matrix is to trace the relationship between categories of requirements. For example:

- Functional requirements and technical requirements
- Security requirements and technical requirements
- Business requirements and technical requirements

An Inter-requirements Traceability Matrix can be used to record this information. A sample form is included after the Requirements Traceability Matrix.

Use the information on your project to tailor the form to best meet your needs.

The Requirements Traceability Matrix can receive information from:

- Project Charter
- Stakeholder Register

It is related to:

- Requirements Management Plan
- Requirements Documentation

It provides information to:

- Product Acceptance
- Change Requests

The Requirements Traceability Matrix is an output from the process 5.1 Collect Requirements in the *PMBOK® Guide*—Fourth Edition.

REQUIREMENTS TRACEABILITY MATRIX

Project Title: _____   Date Prepared: _____

| ID | Requirement Information | | | | | Relationship Traceability | | |
	Requirement	Priority	Category	Source	Relates to Objective	Manifests in WBS Deliverable	Verification	Validation

REQUIREMENTS TRACEABILITY MATRIX

Project Title: _____   Date Prepared: _____

	Requirement Information					Relationship Traceability			
ID	Requirement	Priority	Category	Source	Relates to Objective	Manifests in WBS Deliverable	Verification	Validation
Identifier.	From requirements documentation.	From requirements documentation.	From requirements documentation.	From requirements documentation.	Relationship to objectives in the Project Charter.	Deliverable in the WBS that meets the requirement; can use WBS ID coding.	Method of verifying requirement is met.	Method of validating requirement is met.

INTER-REQUIREMENTS TRACEABILITY MATRIX

Project Title: _____   Date Prepared: _____

ID	Business Requirement	Priority	Source	ID	Technical Requirement	Priority	Source

INTER-REQUIREMENTS TRACEABILITY MATRIX

Project Title: _____   Date Prepared: _____

ID	Business Requirement	Priority	Source	ID	Technical Requirement	Priority	Source
Identifier.	From requirements documentation.	From requirements documentation.	From requirements documentation.	Identifier.	From requirements documentation.	From requirements documentation.	From requirements documentation.

3.5 PROJECT SCOPE STATEMENT

The Project Scope Statement is one of the key documents used to plan the project. It provides information that assists in defining, developing, and constraining the project and product scope. It uses information from the Project Charter and Requirements Documentation and progressively elaborates that information so that deliverables, project exclusions, and acceptance criteria can be defined. The Project Scope Statement is where project constraints and assumptions are documented. Many times the initial assumptions will be documented in the Project Scope Statement and then further elaborated in an Assumption Log. The Project Scope Statement should contain at least this information:

- Product scope description
- Project deliverables
- Product acceptance criteria
- Project exclusions
- Project constraints
- Project assumptions

Use the information from your project to tailor the form to best meet your needs.

The Project Scope Statement can receive information from:

- Project Charter
- Requirements Documentation

It provides information to:

- Work Breakdown Structure
- Network Diagram
- Activity Duration Estimates
- Project Schedule
- Risk Management Plan
- Probability and Impact Assessment
- Project Management Plan

The Project Scope Statement is an output from the process 5.2 Define Scope in the *PMBOK® Guide*—Fourth Edition.

PROJECT SCOPE STATEMENT

Project Title: _____   Date Prepared: _____

Product Scope Description:

Project Deliverables:

Project Acceptance Criteria:

Project Exclusions:

Project Constraints:

Project Assumptions:

PROJECT SCOPE STATEMENT

Project Title: _____   **Date Prepared:** _____

Product Scope Description:

Product scope is progressively elaborated from the project description and the product requirements in the Project Charter.

Project Deliverables:

Project deliverables are progressively elaborated from the project description, the product characteristics, and the product requirements in the Project Charter.

Project Acceptance Criteria:

The acceptance criteria that will need to be met in order for a stakeholder to accept a deliverable. Acceptance criteria can be developed for the entire project or for each component of the project.

Project Exclusions:

Project exclusions clearly define what is considered out of scope for the project.

Project Constraints:

Constraints that may be imposed on the project may include a fixed budget, hard deliverable dates, or specific technology.

Project Assumptions:

Assumptions about deliverables, resources, estimates, and any other aspect of the project that the team holds to be true, real, or correct but has not validated.

3.6 ASSUMPTION AND CONSTRAINT LOG

The Assumption and Constraint Log can be incorporated into the Project Scope Statement or it can be a stand-alone document. Assumptions are factors that, for planning purposes, are considered to be true, real, or certain but without proof or demonstration. This log is a dynamic document since assumptions are progressively elaborated throughout the project. Eventually they are validated and are no longer assumptions. Constraints are an applicable restriction or limitation, either internal or external to a project, that will affect the performance of the project or process. Typical constraints include a predetermined budget or fixed milestones for deliverables. Information in the Assumption and Constraint Log includes:

- Identifier
- Category
- Assumption or constraint
- Responsible party
- Due date
- Actions
- Status
- Comments

Assumptions can come from any document in the project. They can also be determined by the project team. Constraints are generally documented in the Project Charter and are determined by the customer, sponsor, or regulatory agencies.

Although the Assumption and Constraint Log does not explicitly provide information to any specific document, by incorporation in the Project Scope Statement, it provides useful information to:

- Work Breakdown Structure
- Activity Duration Estimates
- Project Schedule
- Risk Management Plan
- Probability and Impact Assessment
- Project Management Plan

It should also be considered when developing Activity Cost Estimates and Activity Resource Requirements.

ASSUMPTION AND CONSTRAINT LOG

Project Title: _____   Date Prepared: _____

ID	Category	Assumption/Constraint	Responsible Party	Due Date	Actions	Status	Comments

ASSUMPTION AND CONSTRAINT LOG

Project Title: _____   Date Prepared: _____

ID	Category	Assumption/Constraint	Responsible Party	Due Date	Actions	Status	Comments
Identifier.	*Area the assumption or constraint impacts.*	*Define the assumption or constraint.*	*Assign assumptions to someone to validate and follow up.*	*Date the assumption should be validated.*	*Any actions needed to validate the assumption or address the constraint.*	*Open, pending, or closed.*	*Any comments to clarify the assumption or constraint or to clarify the status or actions.*

3.7 WORK BREAKDOWN STRUCTURE

The Work Breakdown Structure (WBS) is used to decompose all the work of the project. It begins at the project level and is successively broken down into finer levels of detail. The lowest level is a work package. A work package represents a discrete deliverable that can be decomposed into activities to produce the deliverable. The needs of the project will determine the way that the WBS is organized. The second level determines the organization of the WBS. Some options for organizing and arranging the WBS include:

- Geography
- Major deliverables
- Life cycle
- Subprojects

The WBS should have a method of identifying the hierarchy, such as a numeric structure. The WBS can be shown as a hierarchical chart or as an outline. *The WBS, its corresponding WBS Dictionary, and the Project Scope Statement comprise the scope baseline for the project.*

Use the information from your project to tailor the form to best meet your needs.

The WBS can receive information from:

- Project Scope Statement
- Requirements Documentation

It is related to:

- WBS Dictionary
- Scope Baseline

It provides information to:

- Activity List
- Activity Cost Estimates
- Project Budget
- Quality Management Plan
- Risk Register
- Procurement Management Plan
- Project Management Plan

The WBS is an output from the process 5.3 Create WBS in the *PMBOK® Guide*—Fourth Edition.

WORK BREAKDOWN STRUCTURE

Project Title: _____ Date Prepared: _____

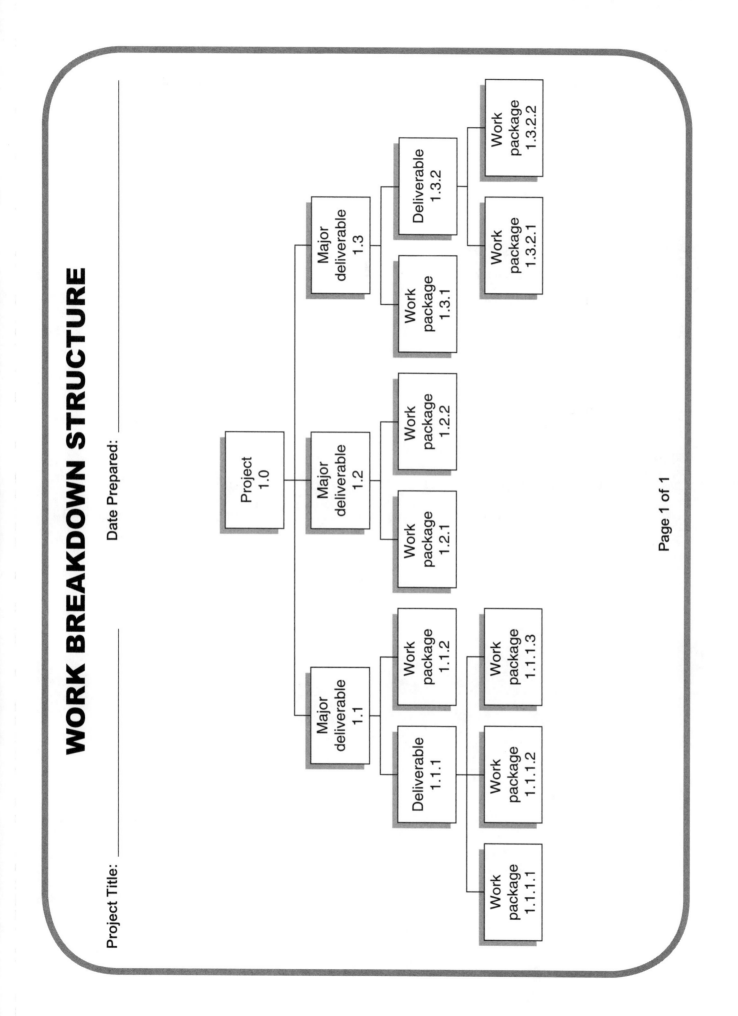

WORK BREAKDOWN STRUCTURE

Project Title: _____   Date Prepared: _____

1. Project

    1.1.   Major deliverable

        1.1.1.   Deliverable

            1.1.1.1.   Work package

            1.1.1.2.   Work package

            1.1.1.3.   Work package

        1.1.2.   Work package

    1.2.   Major deliverable

        1.2.1.   Work package

        1.2.2.   Work package

    1.3.   Major deliverable

        1.3.1.   Work package

        1.3.2.   Deliverable

            1.3.2.1.   Work package

            1.3.2.2.   Work package

3.8 WBS DICTIONARY

The WBS Dictionary supports the Work Breakdown Structure (WBS) by providing detail about the work packages and control accounts it contains. The WBS Dictionary can provide detailed information about each work package or summary information at the control account level and work packages. *The WBS, its corresponding WBS Dictionary, and the Project Scope Statement comprise the scope baseline for the project.* Information in the WBS Dictionary can include:

- WBS identifier
- Description of work
- Responsible organization or person
- List of milestones
- List of schedule activities
- Resources required
- Cost estimates
- Quality requirements
- Acceptance criteria
- Technical information or references
- Contract information

The WBS Dictionary is progressively elaborated as the planning processes progress. Once the WBS is developed, the description of work for a particular work package may be defined, but the necessary activities, cost estimates, and resource requirements may not be known. Thus, the inputs for the WBS Dictionary are more detailed than for the WBS, and there are not as many outputs.

Use the information from your project to tailor the form to best meet your needs.

The WBS Dictionary can receive information from:

- Project Scope Statement
- Requirements Documentation
- Activity List
- Milestone List
- Activity Resource Requirements
- Activity Cost Estimates
- Quality Metrics
- Contracts

It is related to:

- Work Breakdown Structure
- Scope Baseline

It provides information to:

- Risk Register
- Procurement Management Plan

The WBS Dictionary is an output from the process 5.3 Create WBS in the *PMBOK® Guide*—Fourth Edition.

WBS DICTIONARY

Project Title: _____

Date Prepared: _____

Work Package Name: _____ WBS ID: _____

Description of Work:

Milestones:

1.
2.
3.

Due Dates:

ID	Activity	Resource	Labor			Material			Total Cost
			Hours	Rate	Total	Units	Cost	Total	

Quality Requirements:

Acceptance Criteria:

Technical Information:

Contract Information:

WBS DICTIONARY

Project Title: _____ Date Prepared: _____

Work Package Name: *From the WBS* WBS ID: *From the WBS*

Description of Work:
Description of the work to be delivered in sufficient detail to ensure a common understanding by stakeholders.

Milestones: Due Dates:
1. *List any milestones associated with the work package.* *List the due dates of the milestones.*
2.
3.

ID	Activity	Resource	Labor			Material			Total Cost
			Hours	Rate	Total	Units	Cost	Total	
	From activity list or schedule.	*From resource requirements.*	*Total effort.*	*Labor rate.*	*Hours x rate.*	*Amount.*	*Cost.*	*Units x Cost.*	*Labor + Material.*

Quality Requirements:
Quality metrics used to verify the deliverable.

Acceptance Criteria:
Criteria that will be used to accept the WBS element.

Technical Information:
Technical information or reference to technical documentation that contains technical information.

Contract Information:
Relevant contract information that contains constraints, resource information, or other relevant information.

3.9 ACTIVITY LIST

The Activity List defines all the activities necessary to complete the project work. It also describes the work in sufficient detail so that the person performing the work understands the requirements necessary to complete it correctly. The Activity List contains:

- Activity identifier
- Activity name
- Description of work

Use the information from your project to tailor the form to best meet your needs.

The Activity List can receive information from:

- Scope Baseline (particularly the deliverables from the WBS)

It is related to:

- Activity Attributes
- Milestone List

It provides information to:

- Network Diagram
- Activity Resource Requirements
- Activity Duration Estimates
- Gantt Chart or other Schedule

The Activity List is an output from the process 6.1 Define Activities in the *PMBOK® Guide*—Fourth Edition.

ACTIVITY LIST

Project Title: _____   Date Prepared: _____

ID	Activity	Description of Work

ACTIVITY LIST

Project Title: _____   **Date Prepared:** _____

ID	Activity	Description of Work
Follow WBS or schedule numbering scheme.	*Activity name.*	*Description of activity in enough detail so that the person(s) performing the work understands what is required to complete it.*

3.10 ACTIVITY ATTRIBUTES

Activity Attributes are the details about the activity. Sometimes the information is entered directly into the schedule software. Other times the information is collected in a form that can be used later to assist in building the schedule model. Activity Attributes can include:

- Activity identifier or code
- Activity name
- Activity description
- Predecessor and successor activities
- Logical relationships
- Leads and lags
- Imposed dates
- Constraints
- Assumptions
- Required resources and skill levels
- Geographic or location of performance
- Type of effort

The Activity Attributes are progressively elaborated as the planning processes progress. Once the Activity List is complete, the description of work for a particular activity may be defined, but the necessary attributes, such as logical relationships and resource requirements, may not be known. Thus, the inputs for the Activity Attributes are more detailed than for the Activity List.

Use the information from your project to tailor the form to best meet your needs.

The Activity Attributes can receive information from:

- Activity List
- Network Diagram
- Project Scope Statement
- Assumption and Constraint Log
- Activity Resource Requirements

It is related to:

- Milestone List

It provides information to:

- Project Schedule

Activity Attributes are an output from the process 6.1 Define Activities in the *PMBOK® Guide—*Fourth Edition.

ACTIVITY ATTRIBUTES

Project Title: _____   Date Prepared: _____

ID:	Activity:

Description of Work:

Predecessors	Relationship	Lead or Lag	Successor	Relationship	Lead or Lag

Number and Type of Resources Required:	Skill Requirements:	Other Required Resources:

Type of Effort:

Location of Performance:

Imposed Dates or Other Constraints:

Assumptions:

ACTIVITY ATTRIBUTES

Project Title: _____   Date Prepared: _____

ID:	Activity:
From activity list.	From activity list.

Description of Work:

A description of the activity in enough detail so that the person(s) performing the work understands what is required to complete it.

Predecessors	Relationship	Lead or Lag	Successor	Relationship	Lead or Lag
Any activities that must occur before the activity.	The nature of the relationship, such as start-to-start, finish-to-start, or finish-to-finish.	Any required delays between activities (lag) or accelerations (lead).	Any activities that must occur after the activity.	The nature of the relationship, such as start-to-start, finish-to-start, or finish-to-finish.	Any required delays between activities (lag) or accelerations (lead).

Number and Type of Resources Required:

The number and roles of people needed to complete the work.

Skill Requirements:

The level of skill necessary to complete the work (expert, average, novice or applicable job level).

Other Required Resources:

Any equipment, supplies, or other types of resources needed to complete the work.

Type of Effort:

Indicate if the work is a fixed duration, fixed amount of effort, level of effort, apportioned effort or other type of work.

Location of Performance:

If the work is to be completed somewhere other than at the performing organizations site, indicate the location.

Imposed Dates or Other Constraints:

Indicate any fixed delivery dates, milestones or other constraints.

Assumptions:

List any assumptions about resource availability, skill sets, or other assumptions that impact the activity.

3.11 MILESTONE LIST

The Milestone List defines all the project milestones and describes the nature of each one. It may categorize the milestone as optional or mandatory, internal or external, interim or final, or in any other way that supports the needs of the project.

The Milestone List can receive information from:

- Scope Baseline

It is related to:

- Activity List
- Activity Attributes

It provides information to:

- Network Diagram

The Milestone List is an output from the process 6.1 Define Activities in the *PMBOK® Guide*—Fourth Edition.

MILESTONE LIST

Project Title: _____   Date Prepared: _____

Milestone	Milestone Description	Type

MILESTONE LIST

Project Title: _____   Date Prepared: _____

Milestone	Milestone Description	Type
Milestone name.	Description of milestone in enough detail to understand what is needed to meet the milestone.	Internal or external. Interim or final. Mandatory or optional.

3.12 NETWORK DIAGRAM

The Network Diagram is a visual display of the relationship between schedule elements. It can be produced at the activity level, the deliverable level, or the milestone level. The purpose is to visually depict the types of relationships between elements. The elements are shown at nodes that are connected by lines with arrows that indicate the nature of the relationship. Relationships can be of four types:

1. Finish-to-start (FS). This is the most common type of relationship. The predecessor element must be complete before the successor element can begin.
2. Start-to-start (SS). In this relationship, the predecessor element must begin before the successor element begins.
3. Finish-to-finish (FF). In this relationship, the predecessor element must be complete before the successor element can be complete.
4. Start-to-finish (SF). This is the least common type of relationship. The successor element must begin before the predecessor element can be complete.

In addition to the types of relationships, the Network Diagram may show modifications to the relationships, such as leads or lags.

* A lag is a directed delay between elements. In a finish-to-start relationship with a three-day lag, the successor activity would not start until three days after the predecessor was complete. This would be shown as FS+3d. Lag is not float.
* A lead is an acceleration between elements. In a finish-to-start relationship with a three-day lead, the successor activity would begin three days before the predecessor was complete. This would be shown as FS-3d.

Leads and lags can be applied to any type of relationship.
Use the information from your project to determine the level of detail and the need for a Network Diagram.

The Network Diagram can receive information from:

* Project Scope Statement
* Activity List
* Activity Attributes
* Milestone List

It provides information to:

* Project Schedule

The Network Diagram is an output from the process 6.2 Sequence Activities in the *PMBOK® Guide*—Fourth Edition.

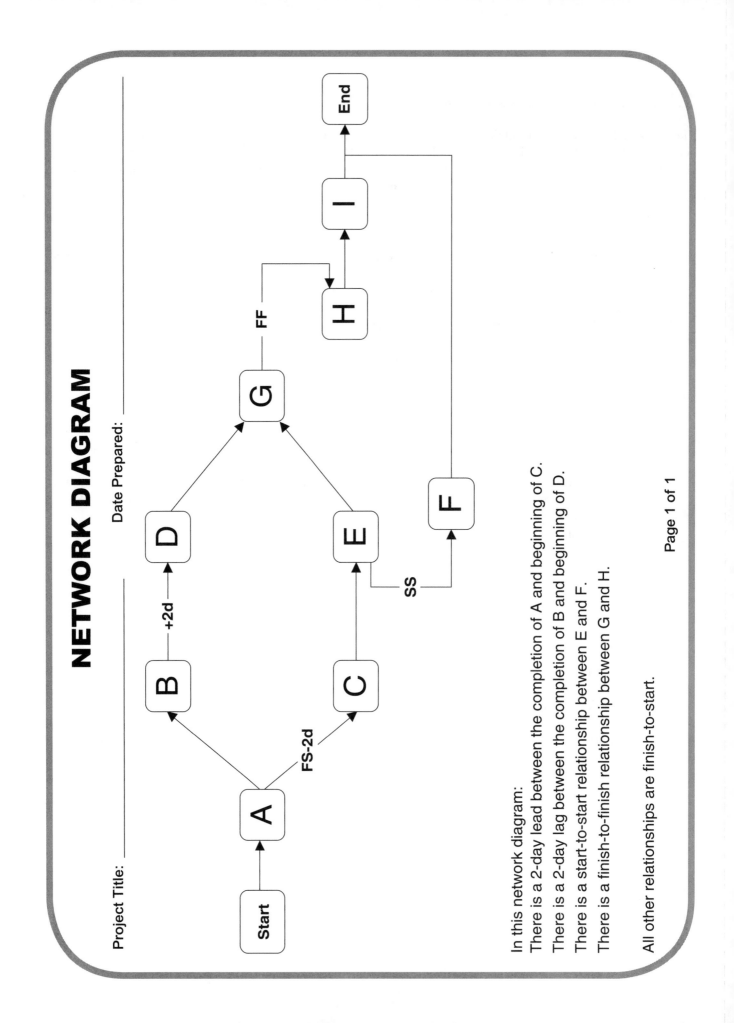

NETWORK DIAGRAM

Project Title: _____   Date Prepared: _____

In this network diagram:

There is a 2-day lead between the completion of A and beginning of C.

There is a 2-day lag between the completion of B and beginning of D.

There is a start-to-start relationship between E and F.

There is a finish-to-finish relationship between G and H.

All other relationships are finish-to-start.

Page 1 of 1

3.13 ACTIVITY RESOURCE REQUIREMENTS

The Activity Resource Requirements describe the type and quantity of resources needed to complete the project work. Resources include:

- People
- Equipment
- Material
- Supplies
- Locations (as needed)

Locations can include training rooms, testing sites, and so on.
Use the information from your project to tailor the form to meet your needs.

The Activity Resource Requirements can receive information from:

- Activity List
- Activity Attributes

It provides information to:

- Resource Breakdown Structure
- Duration Estimating Worksheet
- Project Schedule
- Human Resource Plan
- Procurement Management Plan

Activity Resource Requirements are an output from the process 6.3 Estimate Activity Resources in the *PMBOK® Guide*—Fourth Edition.

ACTIVITY RESOURCE REQUIREMENTS

Project Title: _____   Date Prepared: _____

WBS ID	Type of Resource	Quantity	Comments

Assumptions:

ACTIVITY RESOURCE REQUIREMENTS

Project Title: _____   **Date Prepared:** _____

WBS ID	Type of Resource	Quantity	Comments
From WBS.	*People, equipment, material, supplies, locations, or other.*	*Amount needed.*	*Include special grade, competency, certification, licensure, or other relevant information as needed.*

Assumptions:

Include any assumptions specific to resource requirements.

3.14 RESOURCE BREAKDOWN STRUCTURE

The Resource Breakdown Structure is a hierarchical structure used to organize the resources by type and category. It can be shown as a hierarchical chart or as an outline.

Use the information from your project to tailor the form to best meet your needs.

The Resource Breakdown Structure can receive information from:

* Activity Resource Requirements

The Resource Breakdown Structure is an output from the process 6.3 Estimate Activity Resources in the *PMBOK® Guide*—Fourth Edition.

RESOURCE BREAKDOWN STRUCTURE

Project Title: _____   Date Prepared: _____

1. Project

   1.1. People

      1.1.1. Quantity of Role 1

         1.1.1.1. Quantity of Level 1

         1.1.1.2. Quantity of Level 2

         1.1.1.3. Quantity of Level 3

      1.1.2. Quantity of Role 2

   1.2. Equipment

      1.2.1. Quantity of Type 1

      1.2.2. Quantity of Type 2

   1.3. Materials

      1.3.1. Quantity of Material 1

         1.3.1.1. Quantity of Grade 1

         1.3.1.2. Quantity of Grade 2

   1.4. Supplies

      1.4.1. Quantity of Supply 1

      1.4.2. Quantity of Supply 2

   1.5. Locations

      1.5.1. Location 1

      1.5.2. Location 2

3.15 ACTIVITY DURATION ESTIMATES

Activity Duration Estimates provide information on the amount of time it will take to complete project work. They can be determined by developing an estimate for each work package (called a bottom-up estimate) or by using a quantitative method, such as:

- Parametric estimates
- Analogous estimates
- Three-point estimates

Activity Duration Estimates will generally convert the estimate of effort hours into days or weeks. To convert effort hours into days, take the total number of hours and divide by 8. To convert to weeks, take the total number of hours and divide by 40.

A Duration Estimating Worksheet can assist in developing accurate estimates.

Activity Duration Estimates can receive information from:

- Project Scope Statement
- Activity List
- Activity Attributes
- Duration Estimating Worksheet
- Activity Resource Requirements

They provide information to:

- Project Schedule
- Risk Register

Activity Duration Estimates are an output from the process 6.3 Estimate Activity Resources in the *PMBOK® Guide*—Fourth Edition.

ACTIVITY DURATION ESTIMATES

Project Title: _____   Date Prepared: _____

WBS ID	Activity	Effort Hours	Duration Estimate

ACTIVITY DURATION ESTIMATES

Project Title: _____   Date Prepared: _____

WBS ID	Activity	Effort Hours	Duration Estimate
From WBS.	*Activity name from activity list.*	*400*	*10 weeks*

3.16 DURATION ESTIMATING WORKSHEET

A Duration Estimating Worksheet can help to develop duration estimates when quantitative methods are used. Quantitative methods include:

- Parametric estimates
- Analogous estimates
- Three-point estimates

Parametric estimates are derived by determining the effort hours needed to complete the work. The effort hours are then divided by:

- Resource quantity (i.e., number of people assigned to the task).
- Percent of time the resource(s) are available (i.e., 100 percent of the time, 75 percent of the time, or 50 percent of the time).
- Performance factor. Experts in a field generally complete work faster than people with an average skill level or novices. Therefore, a factor to account for the productivity is developed.

Duration estimates can be made even more accurate by considering that most people are productive on actual work only about 75 percent of the time.

Analogous estimates are derived by comparing current work to previous similar work. The size of the previous work and the duration is compared to the expected size of the current work. Then the size of the current work is multiplied by the previous duration to determine an estimate. Various factors, such as complexity, can be factored in to make the estimate more accurate. This type of estimate is generally used to get a high-level estimate when detailed information is not available.

A three-point estimate can be used to account for uncertainty in the duration estimate. Stakeholders provide estimates for optimistic, most likely, and pessimistic scenarios. These estimates are put into an equation to determine an expected duration. The needs of the project determine the appropriate equation, but a common equation is

$$(\text{optimistic} + 4 \text{ most likely} + \text{pessimistic})/6$$

The Duration Estimating Worksheet can receive information from:

- Project Scope Statement
- Activity List
- Activity Attributes
- Activity Resource Requirements

It provides information to:

- Activity Duration Estimates

DURATION ESTIMATING WORKSHEET

Project Title: _____   Date Prepared: _____

Parametric Estimates

WBS ID	Effort Hours	Resource Quantity	% Available	Performance Factor	Duration Estimate

Analogous Estimates

WBS ID	Previous Activity	Previous Duration	Current Activity	Multiplier	Duration Estimate

Three Point Estimates

WBS ID	Optimistic Duration	Most Likely Duration	Pessimistic Duration	Weighting Equation	Expected Duration Estimate

DURATION ESTIMATING WORKSHEET

Project Title: _____   Date Prepared: _____

Parametric Estimates

WBS ID	Effort Hours	Resource Quantity	% Available	Performance Factor	Duration Estimate
1.1	150	2	0.75	0.8	125

Analogous Estimates

WBS ID	Previous Activity	Previous Duration	Current Activity	Multiplier	Duration Estimate
1.1	Build 160 Sq. ft. deck	10 days	Build 200 Sq. ft. deck	200/160 = 1.25	10 × 1.25 = 12.5 days

Three Point Estimates

WBS ID	Optimistic Duration	Most Likely Duration	Pessimistic Duration	Weighting Equation	Expected Duration Estimate
1.1	20	25	36	$(o + 4m + p)/6$	26

3.17 PROJECT SCHEDULE

The Project Schedule combines the information from the Activity List, Network Diagram, Activity Resource Requirements, Activity Duration Estimates and any other relevant information to determine the start and finish dates for project activities. A common way of showing a schedule is via Gantt chart showing the dependencies between activities. The sample Gantt chart is for designing, building, and installing kitchen cabinets. It shows the:

- WBS identifier
- Activity name
- Start dates
- Finish dates
- Resource name (next to the bar)

The information on your schedule can be much more detailed, depending on the needs of the project. Scheduling software provides many options to record and display information.

Another method of showing schedule information is to create a milestone chart, which shows only the dates of the important events or key deliverables. The sample Milestone Chart is for constructing a house. It shows the activity milestones as well as their dependencies. Showing dependencies on a Milestone Chart is optional.

The Project Schedule can receive information from:

- Project Scope Statement
- Activity List
- Activity Attributes
- Network Diagram
- Activity Resource Requirements
- Activity Duration Estimates

It provides information to:

- Activity Cost Estimates
- Project Budget
- Procurement Management Plan
- Quality Management Plan

The Project Schedule is an output from the process 6.5 Develop Schedule in the *PMBOK® Guide*—Fourth Edition.

PROJECT SCHEDULE

Project Title: _____ Date Prepared: _____

Sample Gantt Chart

ID	WBS	Task Name	Start	Finish
1	**1**	**Kitchen Cabinets**	**Aug 4**	**Oct 2**
2	**1.1**	**Preparation**	**Aug 4**	**Aug 20**
3	1.1.1	Design kitchen layout	Aug 4	Aug 8
4	1.1.2	Design cabinet layout	Aug 6	Aug 12
5	1.1.3	Select materials	Aug 13	Aug 15
6	1.1.4	Purchase materials	Aug 18	Aug 20
7	1.1.5	Preparation complete	Aug 20	Aug 20
8	**1.2**	**Construction**	**Aug 21**	**Sep 26**
9	1.2.1	Build cabinet framing	Aug 21	Sep 10
10	1.2.2	Stain and finish cabinet framing	Sep 11	Sep 12
11	1.2.3	Make cabinet doors	Sep 11	Sep 24
12	1.2.4	Stain and finish doors	Sep 25	Sep 26
13	1.2.5	Make drawers	Sep 11	Sep 17
14	1.2.6	Stain and finish doors	Sep 18	Sep 18
15	1.2.7	Make shelving	Sep 11	Sep 16
16	1.2.8	Stain and finish shelving	Sep 17	Sep 17
17	1.2.9	Construction complete	Sep 26	Sep 26
18	**1.3**	**Installation**	**Sep 29**	**Oct 2**
19	1.3.1	Install cabinet framing	Sep 29	Oct 1
20	1.3.2	Install cabinets	Oct 2	Oct 2
21	1.3.3	Install drawers	Oct 2	Oct 2
22	**1.4**	**Sign off**	Oct 2	Oct 2

PROJECT SCHEDULE

Project Title: _____

Date Prepared: _____

Sample Milestone Chart

ID	❶	Task Name	Finish	Qtr 2, 2008 — Mar Apr May Jun	Qtr 3, 2008 — Jul Aug Sep	Qtr 4, 2008 — Oct Nov Dec	Qtr — Ja
1		Vendors selected	Mar 3	3/3			
2		Financing obtained	Mar 3	3/3			
3	▦	Plans complete	Apr 11	4/11			
4	▦	Permits obtained	May 2	5/2			
5		Paving complete	May 2	5/2			
6	▦	Foundation complete	May 14	5/14			
7	▦	House framed	Jun 13	6/13			
8	▦	Roof set	Jun 20	6/20			
9		Power established	Jun 20	6/20			
10	▦	Power complete	Jul 11	7/11			
11	▦	Plumbing complete	Aug 22		8/22		
12		HVAC complete	Aug 22		8/22		
13	▦	Finish work complete	Sep 26		9/26		
14	▦	Garden site prepared	Oct 10			10/10	
15		City sign-off	Oct 10			10/10	
16	▦	Punch list closed	Oct 17			10/17	

3.18 ACTIVITY COST ESTIMATES

Activity Cost Estimates provide information on the resources necessary to complete project work, including labor, equipment, supplies, services, facilities, and material. Estimates can be determined by developing an approximation for each work package (called a bottom-up estimate) or by using a quantitative method such as:

- Parametric estimates
- Analogous estimates
- Three-point estimates

In addition, information on project reserves, the cost of quality, vendor bids, and indirect costs should be taken into account when developing Activity Cost Estimates.

A Cost Estimating Worksheet can assist in developing accurate estimates.

The cost estimates should provide information on how the estimate was developed, the assumptions and constraints, the range of estimates, and the confidence level.

Activity Cost Estimates can receive information from:

- Scope Baseline
- Project Schedule
- Human Resource Plan
- Cost Estimating Worksheet
- Risk Register

They provide information to:

- Cost Performance Baseline
- Risk Register
- Make-or-buy decisions

Activity Cost Estimates are an output from the process 7.1 Estimate Costs in the *PMBOK® Guide—Fourth Edition*.

ACTIVITY COST ESTIMATES

Project Title: _____   Date Prepared: _____

WBS ID	Resource	Direct Costs	Indirect Costs	Reserve	Estimate	Method	Assumptions/ Constraints	Additional Information	Range	Confidence Level

ACTIVITY COST ESTIMATES

Project Title: _____   Date Prepared: _____

WBS ID	Resource	Direct Costs	Indirect Costs	Reserve	Estimate	Method	Assumptions/ Constraints	Additional Information	Range	Confidence Level
From WBS.	Type of resource, labor, material, etc.	Costs related to the project.	Indirect costs.	Contingency reserve amounts.	Approximate cost.	Method used, such as parametric, analogous, etc.	Any assumptions used in developing the estimate, such as labor cost per hour.	Information on cost of quality, interest rate, or other.	Range of estimate if applicable.	Degree of confidence in the estimate.

3.19 COST ESTIMATING WORKSHEET

A Cost Estimating Worksheet can help to develop cost estimates when quantitative methods or a bottom-up estimates are developed. Quantitative methods include:

* Parametric estimates
* Analogous estimates
* Three-point estimates
* Bottom-up estimates

Parametric estimates are derived by determining the cost variable that will be used and the cost per unit. Then the number of units are multiplied by the cost per unit to derive a cost estimate.

Analogous estimates are derived by comparing current work to previous similar work. The size of the previous work and the cost is compared to the expected size of the current work. Then the size of the current work is multiplied by the previous cost to determine an estimate. Various factors, such as complexity and price increases, can be factored in to make the estimate more accurate. This type of estimate is generally used to get a high-level estimate when detailed information is not available.

A three-point estimate can be used to account for uncertainty in the cost estimate. Stakeholders provide estimates for optimisitic, most likely, and pessimistic scenarios. These estimates are put into an equation to determine an expected cost. The needs of the project determine the appropriate equation, though a common equation is

$$(\text{optimistic} + 4 \text{ most likely} + \text{pessimistic})/6$$

Bottom-up estimates are detailed estimates done at the work-package level. Detailed information on the work package, such as technical requirements, engineering drawings, labor duration and cost estimates, and other direct and indirect costs are used to determine the most accurate estimate possible.

The Cost Estimating Worksheet can receive information from:

* Scope Baseline
* Project Schedule
* Human Resource Plan
* Risk Register

It provides information to:

* Activity Cost Estimates

COST ESTIMATING WORKSHEET

Project Title: _____ Date Prepared: _____

Parametric Estimates

WBS ID	Cost Variable	Cost per Unit	Number of Units	Cost Estimate

Analogous Estimates

WBS ID	Previous Activity	Previous Cost	Current Activity	Multiplier	Cost Estimate

Three Point Estimates

WBS ID	Optimistic Cost	Most Likely Cost	Pessimistic Cost	Weighting Equation	Expected Cost Estimate

COST ESTIMATING WORKSHEET

Project Title: _____   Date Prepared: _____

Parametric Estimates

WBS ID	Cost Variable	Cost per Unit	Number of Units	Cost Estimate
1.1	Square feet	$9.50	36	$342

Analogous Estimates

WBS ID	Previous Activity	Previous Cost	Current Activity	Multiplier	Cost Estimate
1.1	Build 160 sq. ft. deck	$5,000	Build 200 sq. ft. deck	1.25	$6,250

Three Point Estimates

WBS ID	Optimistic Cost	Most Likely Cost	Pessimistic Cost	Weighting Equation	Expected Cost Estimate
1.1	$4,000	$5,000	$7,500	$(o + 4m + p)/6$	$5,250

BOTTOM-UP COST ESTIMATING WORKSHEET

Project Title: _____   Date Prepared: _____

WBS ID	Labor Hours	Labor Rates	Total Labor	Material	Supplies	Equipment	Travel	Other Direct Costs	Indirect Costs	Reserve	Estimate

BOTTOM-UP COST ESTIMATING WORKSHEET

Project Title: _____   Date Prepared: _____

WBS ID	Labor Hours	Labor Rates	Total Labor	Material	Supplies	Equipment	Travel	Other Direct Costs	Indirect Costs	Reserve	Estimate
From WBS.	*From duration estimates.*	*By hour, day or fixed rate.*	*Hours x rates.*	*From quotes.*	*From quotes.*	*From quotes.*	*From quotes.*	*As appropriate.*	*Per company policy.*	*As appropriate.*	*Sum of all costs.*

3.20 COST PERFORMANCE BASELINE

The Cost Performance Baseline is a time-phased budget that is used to measure, monitor, and control cost performance for the project. A project may have multiple performance baselines; for example, the project manager may keep a separate baseline for labor or procurements. The baseline may or may not include contingency funds or indirect costs. When earned value measurements are being used, the baseline may be called the performance measurement baseline.

The needs of the project will determine the information that should be used in the Cost Performance Baseline.

The Cost Performance Baseline can receive information from:

* Scope Baseline
* Project Schedule
* Contracts
* Activity Cost Estimates

It provides information to:

* Make-or-buy decisions
* Project Management Plan
* Quality Management Plan

The Cost Performance Baseline is an output from the process 7.2 Determine Budget in the *PMBOK® Guide—Fourth Edition*.

COST PERFORMANCE BASELINE

Project Title: _____   Date Prepared: _____

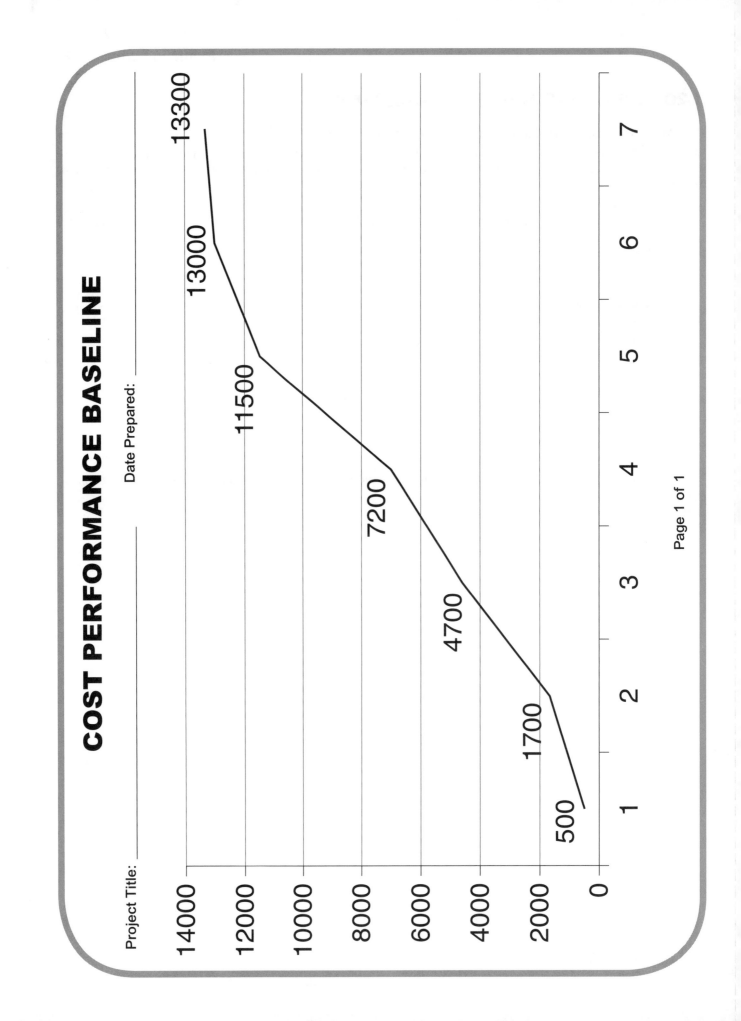

3.21 QUALITY MANAGEMENT PLAN

The Quality Management Plan is a component of the Project Management Plan. It describes how quality requirements for the project will be met. Information in the Quality Management Plan can include:

* Roles and responsibilities
* Quality assurance approach
* Quality control approach
* Quality improvement approach

It may also define the tools, processes, policies, and procedures that will be used to implement the plan. Some projects may combine the Quality Management Plan with the Process Improvement Plan and the Quality Metrics or quality checklist. Other projects will have a separate document for each.

Use the information from your project to tailor the form to best meet your needs.

The Quality Management Plan can receive information from:

* Scope Baseline
* Schedule Baseline
* Cost Performance Baseline
* Stakeholder Register
* Risk Register

It is related to:

* Quality Metrics
* Process Improvement Plan

It provides information to:

* Project Management Plan
* Risk Register

The Quality Management Plan is an output from the process 8.1 Plan Quality in the *PMBOK® Guide*—Fourth Edition.

QUALITY MANAGEMENT PLAN

Project Title: _____   **Date Prepared:** _____

Quality Roles and Responsibilities:

Role:	Responsibilities:
1.	1.
2.	2.
3.	3.
4.	4.

Quality Assurance Approach:

Quality Control Approach:

Quality Improvement Approach:

QUALITY MANAGEMENT PLAN

Project Title: _____ **Date Prepared:** _____

Quality Roles and Responsibilities:

Role:	Responsibilities:
1. *Describe the role needed.*	1. *Describe the responsibilities associated with the role.*
2.	2.
3.	3.
4.	4.

Quality Assurance Approach:

Describe the processes, procedures, methods, tools, and techniques that will be used in performing quality assurance activities.

Quality Control Approach:

Describe the processes, procedures, methods, tools, and techniques that will be used in performing quality control activities.

Quality Improvement Approach:

Describe the processes, procedures, methods, tools, and techniques that will be used in performing quality improvement activities.

3.22 QUALITY METRICS

Quality Metrics provide detailed specific measurements about a project, product, service, or result and how it should be measured. Metrics are consulted in the quality assurance process to ensure that the processes used will meet the metric. The deliverables or processes are measured in the quality control process and compared to the metric to determine if the result is acceptable or if corrective action or rework is required.

The needs of the project will determine the appropriate metrics.

Quality Metrics can receive information from:

- Scope Baseline
- Schedule Baseline
- Cost Performance Baseline
- Stakeholder Register
- Risk Register

They are related to:

- Quality Management Plan
- Process Improvement Plan

Quality Metrics are an output from the process 8.1 Plan Quality in the *PMBOK® Guide*—Fourth Edition.

QUALITY METRICS

Project Title: _____   Date Prepared: _____

ID	Item	Metric	Measurement Method

QUALITY METRICS

Project Title: _____   Date Prepared: _____

ID	Item	Metric	Measurement Method
WBS or other identifier.	*Item to be measured.*	*Measurement.*	*Method of measuring.*

3.23 PROCESS IMPROVEMENT PLAN

The Process Improvement Plan is a component of the Project Management Plan. It describes the approach to selecting and analyzing work processes that can be improved. The processes can be project management specific, project specific, or in the case of a process improvement project, they can be organization wide. The Process Improvement Plan can include:

* Description of processes for improvement
* Flowchart of the process including its inputs, outputs, and interfaces
* Process metrics (if not in the quality metric form)
* Targets for improvement
* Approach for improvement

It may also define the tools, processes, policies, and procedures that will be used to implement the plan. Use the information from your project to tailor the form to best meet your needs.

The Process Improvement Plan can receive information from:

* Scope Baseline
* Stakeholder Register

It is related to:

* Quality Management Plan
* Quality Metrics

It provides information to:

* Project Management Plan

The Process Improvement Plan is an output from the process 8.1 Plan Quality in the *PMBOK® Guide—Fourth Edition.*

PROCESS IMPROVEMENT PLAN

Project Title: _____   Date Prepared: _____

Process Description:

\
\
\
\
\

Process Metrics:

\
\
\
\
\
\
\

Targets for Improvement:

\
\
\
\
\
\

Process Improvement Approach:

\
\
\
\
\
\

Attach a process flowchart of the current and the intended future processes.

PROCESS IMPROVEMENT PLAN

Project Title: _____   Date Prepared: _____

Process Description:

A description of the process including the start and end of the process, the stakeholders of the process, and the inputs, outputs and interfaces of the process. The stakeholders can be end users, maintenance and operations, or machines and equipment. Any other relevant information about the process should be included to provide sufficient understanding.

Process Metrics:

The metrics and measurements involved in the process. This can include time, number or steps or hand-offs, current errors, etc. The metrics in this section represent the current process, not the improved process. This is sometimes called the "as-is" process.

Targets for Improvement:

An explicit statement of the aspect of the process targeted for improvement and the intended metrics. This is sometimes called the "to-be" process.

Process Improvement Approach:

A description of the skills, processes, approaches, tools, and techniques that will be applied to improve the process.

Attach a process flowchart of the current and the intended future processes.

3.24 RESPONSIBILITY ASSIGNMENT MATRIX

The Responsibility Assignment Matrix (RAM) shows the intersection of work packages and resources. Generally RAMs are used to show the different levels of participation on a work package by various team members, but they can also show equipment and materials can be used on work packages. RAMs can indicate different types of participation depending on the needs of the project. Some common types include:

* Accountable
* Responsible
* Consulted
* Resource
* Informed
* Sign-off

The RAM should include a key that explains what each of the levels of participation entails. The next page shows an example using a RACI chart, as demonstrated in the *PMBOK® Guide*—Fourth Edition. The needs of your project should determine the fields for the RAM you use.

The Responsibility Assignment Matrix can receive information from:

* Activity Resource Requirements

It is related to:

* Roles and Responsibilities
* Human Resource Plan

The Responsibility Assignment Matrix is a technique used in the process 9.1 Develop Human Resource Plan in the *PMBOK® Guide*—Fourth Edition.

RESPONSIBILITY ASSIGNMENT MATRIX

Project Title: _____   **Date Prepared:** _____

	Person 1	Person 2	Person 3	Person 4	Etc.
Work package 1	R	C	A		
Work package 2		A		I	R
Work package 3		R	R	A	
Work package 4	A	R	I	C	
Work package 5	C	R	R		A
Work package 6	R		A	I	
Etc.	C	A		R	R

R = Responsible: The person performing the work.

A = Accountable: The person who is answerable to the project manager that the work is done on time, meets requirements and is acceptable.

C = Consult: This person has information necessary to complete the work.

I = Inform: This person should be notified when the work is complete.

3.25 ROLES AND RESPONSIBILITIES

Roles and Responsibilities describe the attributes of a position on the project team. Some common attributes include:

* Responsibility
* Authority
* Qualifications
* Competencies

The resource Roles and Responsibilities description can receive information from:

* Activity Resource Requirements

It is related to:

* Responsibility Assignment Matrix
* Human Resource Plan

Roles and Responsibilities are a component of the Human Resource Plan which is an output from the process 9.1 Develop Human Resource Plan in the *PMBOK® Guide*—Fourth Edition.

ROLES AND RESPONSIBILITIES

Project Title: _____   Date Prepared: _____

Resource Role Description:

[ ]

Authority:

[ ]

Responsibility:

[ ]

Qualifications:

[ ]

Competencies:

[ ]

ROLES AND RESPONSIBILITIES

Project Title: _____ Date Prepared: _____

Resource Role Description:

Provides the role or job title and a brief description of the role.

Authority:

Defines the decision-making limits for the role. Examples include alternative selection, conflict management, prioritizing, rewarding and penalizing, etc. Also indicates reporting structure.

Responsibility:

Defines the activities that the role carries out and the nature of the contribution to the final product, service, or result. Examples include job duties, processes involved, and hand-offs to other roles.

Qualifications:

Describes any prerequisites, experience, licenses, seniority levels, or other qualifications necessary to fulfill the role.

Competencies:

Describes specific role or job skills and competencies. May include details on languages, technology, or other information necessary to complete the role successfully.

3.26 HUMAN RESOURCE PLAN

The Human Resource Plan is part of the Project Management Plan. It describes how all aspects of human resources should be addressed. It is composed of at least three sections:

1. Roles and Responsibilities
2. Project organization charts
3. Staffing Management Plan

The Roles and Responsibilities section uses the information from the Roles and Responsibilities form. That form can be fully incorporated into the Human Resource Plan as is, or information on the roles, authority, responsibilities, qualifications, and requirements can be entered separately.

The project organizational charts can be presented in a graphic hierarchical structure or an outline form. The charts should show the structure within the project, how the project fits in the overall organization, and any dotted-line reporting with the rest of the organization.

The Staffing Management Plan includes information on how the human resource requirements will be met. It includes information on such topics as:

* Staff acquisition
* Staff release
* Resource calendars
* Training needs
* Rewards and recognition
* Regulation, standard, and policy compliance
* Safety

The Human Resource Plan can receive information from:

* Activity Resource Requirements

It is related to:

* Responsibility Assignment Matrix
* Roles and Responsibilities

It provides information to:

* Project Management Plan
* Activity Cost Estimates

The Human Resource Plan is an output from the process 9.1 Develop Human Resource Plan in the *PMBOK® Guide*—Fourth Edition.

HUMAN RESOURCE PLAN

Project Title: _____ Date Prepared: _____

Roles, Responsibilities, and Authority:

Role:	Authority:	Responsibility:
1.	1.	1.
2.	2.	2.
3.	3.	3.
4.	4.	4.
5.	5.	5.
6.	6.	6.

Project Organizational Structure:

HUMAN RESOURCE PLAN

Staffing Management Plan

Staff Acquisition:

Staff Release:

Resource Calendars:

Training Needs:

Rewards and Recognition:

Regulations, Standards, and Policy Compliance:

Safety:

HUMAN RESOURCE PLAN

Project Title: _____ Date Prepared: _____

Roles, Responsibilities and Authority:

Role:	Authority:	Responsibility:
1. *Defines the role or job title.*	1. *Defines decision making limits.*	1. *Defines the duties.*
2.	2.	2.
3.	3.	3.
4.	4.	4.
5.	5.	5.
6.	6.	6.

Project Organizational Structure:

Insert an organizational chart for the project.
May include an organizational chart of the enterprise and how the project fits within the enterprise.

HUMAN RESOURCE PLAN

Staffing Management Plan

Staff Acquisition:

Describes how staff will be brought on to the project. Defines any differences between internal staff team members and outsourced team members with regards to on boarding procedures.

Staff Release:

Describes how team members will be released from the team, including knowledge transfer, and check out procedures for staff and outsourced team members.

Resource Calendars:

Shows any unusual resource calendars such as abbreviated work weeks, vacations, and time constraints for team members that are less than full time.

Training Needs:

Describes any required training on equipment, technology or company processes.

Rewards and Recognition:

Describes any reward and recognition processes or limitations.

Regulations, Standards and Policy Compliance:

Describes any regulations, standards or policies that must be used and how compliance will be demonstrated.

Safety:

Describes any safety regulations, equipment, training or procedures.

3.27 COMMUNICATIONS MANAGEMENT PLAN

The Communications Management Plan is a component of the Project Management Plan. It describes the communications needs of the project including audiences, messages, methods, and other relevant information. Typical information includes:

- Message to be communicated
- Audience
- Media or method
- Frequency
- Sender

In addition, the Communications Management Plan can include a glossary of project-specific terms, flowcharts of how information moves, constraints and assumptions, and methods for addressing sensitive or proprietary information.

The Communications Management Plan can receive information from:

- Stakeholder Register
- Stakeholder Management Strategy

It provides information to:

- Project Management Plan
- Risk Management Plan

The Communications Management Plan is an output from the process 10.1 Plan Communications in the *PMBOK® Guide—Fourth Edition*.

COMMUNICATIONS MANAGEMENT PLAN

Project Title: _____ Date Prepared: _____

Message	Audience	Method	Frequency	Sender

Term or Acronym	Definition

Communication Constraints or Assumptions:

Attach relevant communication diagrams or flowcharts.

COMMUNICATIONS MANAGEMENT PLAN

Project Title: _____ Date Prepared: _____

Message	Audience	Method	Frequency	Sender
Describe the information to be communicated: For example, status reports, project updates, meeting minutes, etc.	List the people or the groups of people who should receive the information.	Describe how the information will be delivered. For example, e-mail, meetings, Web meetings, etc.	List how often the information is to be provided.	Insert the name of the person or the group that will provide the information.

Term or Acronym	Definition
List any terms or acronyms unique to the project or that are used in a unique way.	Provide a definition of the term or the full term for acronyms.

Communication Constraints or Assumptions:

List any assumptions or constraints. Constraints can include descriptions of proprietary information and relevant restrictions on distribution.

Attach relevant communication diagrams or flowcharts.

3.28 RISK MANAGEMENT PLAN

The Risk Management Plan is a component of the Project Management Plan. It describes the approach for managing uncertainty, both threats and opportunities, for the project. Typical information includes:

- Methods and approaches
- Tools and techniques used in risk management
- Roles and responsibilities for risk management
- Categories of risks
- Stakeholder tolerance information
- Definitions of probability
- Definitions of impact by objective
- Probability and impact matrix template
- Funds needed to identify, analyze, and respond to risk
- Protocols for establishing budget and schedule contingency
- Frequency and timing of risk activities
- Risk audit approaches

Not all projects need to this level of detail. Use the information from your project to tailor the form to best meet your needs.

The Risk Management Plan can receive information from:

- Project Scope Statement
- Schedule Management Plan
- Cost Management Plan
- Communications Management Plan

It provides information to:

- Project Management Plan
- Risk Register

The Risk Management Plan is an input to all the other planning processes in the risk management knowledge area. The Risk Register is the only document that is a discrete output from these processes. The Risk Management Plan provides key information needed to conduct those processes successfully.

The Risk Management Plan is an output from the process 11.1 Plan Risk Management in the *PMBOK*®
Guide—Fourth Edition.

RISK MANAGEMENT PLAN

Project Title: _____ Date Prepared: _____

Methods and Approaches:

[ ]

Tools and Techniques:

[ ]

Roles and Responsibilities:

[ ]

Risk Categories:

[ ]

Stakeholder Risk Tolerance:

[ ]

RISK MANAGEMENT PLAN

Definitions of Probability:

Definitions of Impact by Objective:

Probability and Impact Matrix:

RISK MANAGEMENT PLAN

Risk Management Funding:

```
┌──────────────────────────────────────────────────┐
│                                                    │
│                                                    │
│                                                    │
│                                                    │
│                                                    │
└──────────────────────────────────────────────────┘
```

Contingency Protocols:

```
┌──────────────────────────────────────────────────┐
│                                                    │
│                                                    │
│                                                    │
│                                                    │
│                                                    │
│                                                    │
└──────────────────────────────────────────────────┘
```

Frequency and Timing:

```
┌──────────────────────────────────────────────────┐
│                                                    │
│                                                    │
│                                                    │
│                                                    │
│                                                    │
└──────────────────────────────────────────────────┘
```

Risk Audit Approach:

```
┌──────────────────────────────────────────────────┐
│                                                    │
│                                                    │
│                                                    │
│                                                    │
│                                                    │
└──────────────────────────────────────────────────┘
```

RISK MANAGEMENT PLAN

Project Title: _____ Date Prepared: _____

Methods and Approaches:

Describe the methodology or approach to risk management. Provide information on how each of the risk management processes will be carried out, including whether quantitative risk analysis will be performed and under what circumstances.

Tools and Techniques:

Describe the tools, such as a risk breakdown structure, and techniques, such as interviewing, Delphi technique, etc., that will be used for each process.

Roles and Responsibilities:

Describe the roles and responsibilities for various risk management activities.

Risk Categories:

Identify any categorization groups used to sort and organize risks. These can be used to sort risks on the risk register or for a risk breakdown structure, if one is used.

Stakeholder Risk Tolerance:

Describe the risk tolerance levels of the organization(s) and key stakeholders on the project.

RISK MANAGEMENT PLAN

Definitions of Probability:

Terms used to measure probability, such as Very Low–Very High, or .01–1.0.	*Describe the ways of measuring probability: the difference between very high and high probability, etc.. If using a numeric scale, identify the spread between bands of probability (.05, .1, .2, .4, .8 or .2, .4, .6, .8).*

Definitions of Impact by Objective:

Impact	Scope	Quality	Schedule	Cost
Specify terms used to measure impact, such as Very Low–Very High, or.01–1.0.	*Describe the ways of measuring impact on each objective. Objectives other than the ones listed here can be used. Define the difference between very high and high impact on an objective. If using a numeric scale, identify the spread between bands of impact (.05, .1, .2, .4, .8 or .2, .4, .6, .8). Note that the impacts on individual objectives may be different if one objective is more important than another.*			

Probability and Impact Matrix*:

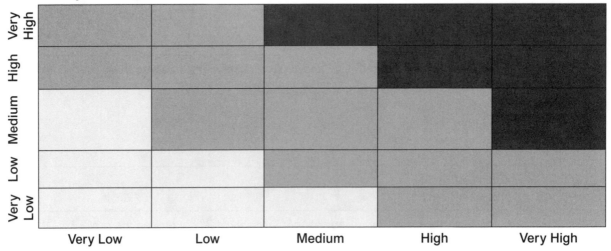

*This is a sample matrix for one project objective. The shading shows a balanced matrix that indicates ranking of High, Medium, or Low based on the probability and impact scores. The darkest shade indicates high risks, the mid-shade indicates medium risks, and the light shade is for low risks.

RISK MANAGEMENT PLAN

Risk Management Funding:

Define the funding needed to perform the various risk management activities, such as utilizing expert advice or transferring risks to a third party.

Contingency Protocols:

Describe the guidelines for establishing, measuring, and allocating both budget contingency and schedule contingency.

Frequency and Timing:

Describe the frequency of conducting formal risk management activities and the timing of any specific activities.

Risk Audit Approach:

Describe how often the risk management process will be audited, which aspects will be audited, and how discrepancies will be addressed.

3.29 RISK REGISTER

The Risk Register is used to track information about identified risks over the course of the project. Typical information includes:

- Risk identifier
- Risk statement
- Probability of occuring
- Impact on objectives if the risk occurs
- Risk score
- Response strategies
- Revised probability
- Revised impact
- Revised score
- Responsible party
- Actions
- Status
- Comments

Not all projects need this level of detail.

Use the information from your project to tailor the Risk Register to best meet your needs. The Risk Register can receive information from anywhere in the project environment. Some documents that should be specifically reviewed for input include:

- Risk Management Plan
- Scope Baseline
- Activity Duration Estimates
- Activity Cost Estimates
- Quality Management Plan
- Stakeholder Register

It provides information to:

- Activity Cost Estimates
- Quality Management Plan
- Procurement Management Plan

The Risk Register is an output from the process 11.2 Identify Risks in the *PMBOK® Guide*—Fourth Edition.

RISK REGISTER

Project Title: _____

Date Prepared: _____

Risk ID	Risk Statement	Probability	Impact				Score	Response
			Scope	Quality	Schedule	Cost		

Revised Probability	Revised Impact				Revised Score	Responsible Party	Actions	Status	Comments
	Scope	Quality	Schedule	Cost					

RISK REGISTER

Project Title: _____   Date Prepared: _____

Risk ID	Risk Statement	Probability	Impact				Score	Response
			Scope	Quality	Schedule	Cost		
Identifier.	Description of the risk event or circumstance.	Likelihood of occurrence.	Impact on each objective if it does occur.				Probability × impact.	Description of planned response strategy to the risk event.

Revised Probability	Revised Impact				Revised Score	Responsible Party	Actions	Status	Comments
	Scope	Quality	Schedule	Cost					
Likelihood after the response strategy.	Revised impact on each objective after the response strategy.				Revised probability × impact.	Who will follow through on the risk and response.	Actions that need to be taken to address the risk.	Open or closed.	Any comments that provide information about the risk.

3.30 PROBABILITY AND IMPACT ASSESSMENT

The Probability and Impact Assessment contains narrative descriptions of the likelihood of events occurring and the impact on the various project objectives if they do occur. It also has a key to assign an overall risk rating based on the probability and impact scores. If a Risk Management Plan is used, this information will become part of that plan. If a Risk Management Plan is not used, this form defines how risks will be analyzed.

The sample forms show descriptions for scope, quality, schedule, and cost objectives. Some projects also rate stakeholder satisfaction as an objective. On smaller projects, the impacts may be grouped together without distinguishing impact by objective. Your project should determine the objectives that are used.

The sample forms use a scale of Very Low to Very High. Some projects use a scale of 1 to 3 or 1 to 5 or percentages. As long as there is a consistent understanding of the rating and ranking system, either approach is acceptable.

Many projects prioritize project objectives. In this case, the impact scale may become more conservative for those objectives that are considered most important. In such cases the probability, impact, and risk rating may all reflect the relative importance of objectives. Another aspect of risk assessment is the urgency of a risk event. Some scales rate the additional variable of urgency to indicate whether the event is imminent or in the distant future.

Use the information from your project to tailor the assessment levels to best meet your needs.

Information in this form provides information to:

* Probability and Impact Risk Matrix
* Risk Register

The Probability and Impact Assessment is a tool used in the process 11.3 Perform Qualitative Risk Analysis in the *PMBOK® Guide*—Fourth Edition.

PROBABILITY AND IMPACT ASSESSMENT

Project Title: _____ **Date Prepared:** _____

Scope Impact:

Very High	
High	
Medium	
Low	
Very Low	

Quality Impact:

Very High	
High	
Medium	
Low	
Very Low	

Schedule Impact:

Very High	
High	
Medium	
Low	
Very Low	

Cost Impact:

Very High	
High	
Medium	
Low	
Very Low	

PROBABILITY AND IMPACT ASSESSMENT

Probability:

Very High	
High	
Medium	
Low	
Very Low	

Risk Rating:

High	
Medium	
Low	

PROBABILITY AND IMPACT ASSESSMENT

Project Title: _____   Date Prepared: _____

Scope Impact:

Very High	The product does not meet the objectives and is effectively useless.
High	The product is deficient in multiple essential requirements.
Medium	The product is deficient in one major requirement or multiple minor requirements.
Low	The product is deficient in a few minor requirements.
Very Low	Minimal deviation from requirements.

Quality Impact:

Very High	Performance is significantly below objectives and is effectively useless.
High	Major aspects of performance do not meet requirements.
Medium	At least one performance requirement is significantly deficient.
Low	There is minor deviation in performance.
Very Low	Minimal deviation in performance.

Schedule Impact:

Very High	Greater than 20% overall schedule increase.
High	Between 10% and 20% overall schedule increase.
Medium	Between 5% and 10% overall schedule increase.
Low	Noncritical paths have used all their float, or overall schedule increase of 1% to 5%.
Very Low	Slippage on noncritical paths but float remains.

Cost Impact:

Very High	Cost increase of greater than 20%.
High	Cost increase of 10% to 20%.
Medium	Cost increase of 5% to 10%.
Low	Cost increase that requires use of all contingency funds.
Very Low	Cost increase that requires use of some contingency but some contingency funds remain.

PROBABILITY AND IMPACT ASSESSMENT

Probability:

Very High	The event will most likely occur: 80% or greater probability.
High	The event will probably occur: 61% to 80% probability.
Medium	The event is likely to occur: 41% to 60% probability.
Low	The event may occur: 21% to 40% probability.
Very Low	The event is unlikely to occur: 1% to 20% probability.

Risk Rating:

High	Any event with a probability of medium or above and a very high impact on any objective.
	Any event with a probability of high or above and a high impact on any objective.
	Any event with a probability of very high and a medium impact on any objective.
	Any event that scores a medium on more than two objectives.
Medium	Any event with a probability of very low and a high or above impact on any objective.
	Any event with a probability of low and a medium or above impact on any objective.
	Any event with a probability of medium and a low to high impact on any objective.
	Any event with a probability of high and a very low to medium impact on any objective.
	Any event with a probability of very high and a low or very low impact on any objective.
	Any event with a probability of very low and a medium impact on more than two objectives.
Low	Any event with a probability of medium and a very low impact on any objective.
	Any event with a probability of low and a low or very low impact on any objective.
	Any event with a probability of very low and a medium or less impact on any objective.

3.31 PROBABILITY AND IMPACT MATRIX

The Probability and Impact Matrix is a table that is used to plot each risk after performing a risk assessment. The Probability and Impact Assessment determines the probability and impact of the risk. This matrix provides a helpful way to look at the various risks on the project and prioritize them for responses. It also provides an overview of the amount of risk on the project. The project team can get an idea of the overall project risk by seeing the number of risks in each square of the matrix. A project with many risks in the red zone will need more contingency to absorb the risk and likely more time and budget to develop and implement risk responses. In some situations a decision is made not to pursue a project because there is more risk than the organization is willing to absorb.

A sample is on the next page. The needs of your project will determine the exact lay out of the matrix.

The Probability and Impact Matrix can receive information from:

* Risk Register
* Probability and Impact Assessment

It provides additional information to the Risk Register.

The Probability and Impact Matrix is a tool used in the process 11.3 Perform Qualitative Risk Analysis in the *PMBOK® Guide*—Fourth Edition.

PROBABILITY AND IMPACT MATRIX

Project Title: _____   Date Prepared: _____

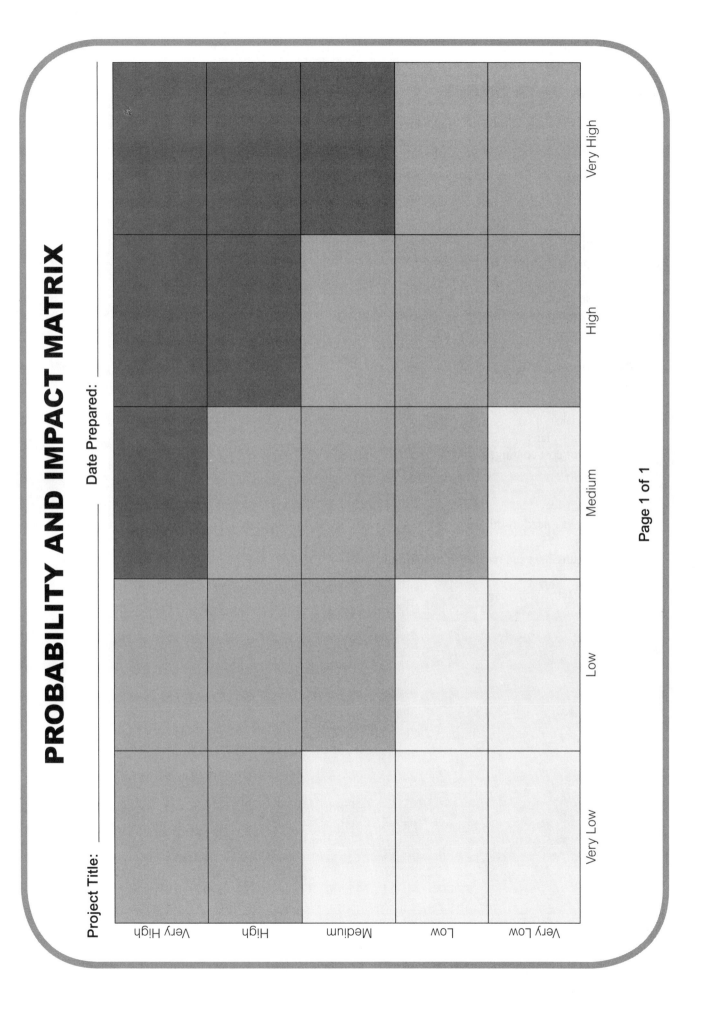

Page 1 of 1

3.32 RISK DATA SHEET

A Risk Data Sheet contains information about a specific identified risk. The information is filled in from the Risk Register and updated with more detailed information. Typical information includes:

- Risk identifier
- Risk description
- Status
- Risk cause
- Probability
- Impact on each objective
- Risk score
- Response strategies
- Revised probability
- Revised impact
- Revised score
- Responsible party
- Actions
- Secondary risks
- Residual risks
- Contingency plans
- Schedule or cost contingency
- Fallback plans
- Comments

Not all projects need to this level of detail. Use the information from your project to tailor the form to best meet your needs.

The Risk Data Sheet can receive information from:

- Risk Register

RISK DATA SHEET

Project Title: _____   Date Prepared: _____

Risk ID:

Risk Description:

Status:

Risk Cause:

	Impact			Score
Probability	Scope	Quality	Schedule	Cost

	Revised Impact			Revised Score
Revised Probability	Scope	Quality	Schedule	Cost

Responses		
Responsible Party		Actions

Secondary Risks:

Residual Risk:

Contingency Plan: Contingency Funds:
                                                            Contingency Time:

Fallback Plans:

Comments:

RISK DATA SHEET

Project Title: _____   Date Prepared: _____

Risk ID:
Risk identifier.

Risk Description:
Detailed description of the risk.

Status:
Open or closed.

Risk Cause:
Description of the circumstances or drivers that are the source of the risk.

Probability	Impact			Score	Responses
	Scope	Quality	Schedule	Cost		
Qualitative or quantitative.	Qualitative or quantitative assessment of the impact on each objective.				Probability x impact.	Response strategies for the event. Use multiple strategies where appropriate.

Revised Probability	Revised Impact			Revised Score	Responsible Party	Actions
	Scope	Quality	Schedule	Cost		
Qualitative or quantitative.	Qualitative or quantitative assessment of the impact on each objective.				Probability x impact.	Person who will manage the risk.	Actions needed to implement responses.

Secondary Risks:
Description of new risks that arise out of the response strategies taken to address the risk.

Residual Risk:
Description of the remaining risk after response strategies.

Contingency Plan:
A plan that will be initiated if specific events occur, such as missing an intermediate milestone. Contingency plans are used when the risk or residual risk is accepted.

Contingency Funds:
Funds needed to protect the budget from overrun.

Contingency Time:
Time needed to protect the schedule from overrun.

Fallback Plans:
A plan devised for use if other response strategies fail.

Comments:
Any other information on the risk, the status of the risk, or response strategies.

3.33 PROCUREMENT MANAGEMENT PLAN

The Procurement Management Plan is a component of the Project Management Plan. It describes how all aspects of a procurement will be managed. The plan provides information to the other processes in the Project Procurement Management knowledge area. Typical information includes:

- Procurement roles and responsibility
- Standard procurement documents
- Contract type
- Statement of work requirements
- Prequalified seller lists
- Bonding and insurance requirements
- Selection criteria
- Procurement constraints and assumptions

Procurements must be integrated with the rest of the project work. Additional information on integration can include:

- WBS integration requirements
- Schedule integration requirements, including lead times and milestones
- Documentation requirements
- Risk management requirements
- Performance reporting requirements

The Procurement Management Plan can receive information from:

- Risk Register
- Requirements Documentation
- Scope Baseline
- Activity Resource Requirements
- Project Schedule
- Activity Cost Estimates
- Cost Performance Baseline

It provides information to:

- Project Management Plan
- Stakeholder Register

The Procurement Management Plan is an output from the process 12.1 Plan Procurements in the *PMBOK® Guide*—Fourth Edition.

PROCUREMENT MANAGEMENT PLAN

Project Title: _____   Date Prepared: _____

Procurement Authority:

Roles and Responsibilities:

Project Manager:	Procurement Department:
1.	1.
2.	2.
3.	3.
4.	4.
5.	5.

Standard Procurement Documents:

1.
2.
3.
4.
5.

Contract Type:

Bonding and Insurance Requirements:

Selection Criteria:

Weight	Criteria

Procurement Assumptions and Constraints:

PROCUREMENT MANAGEMENT PLAN

Integration Requirements:

WBS	
Schedule	
Documentation	
Risk	
Performance Reporting	

PROCUREMENT MANAGEMENT PLAN

Project Title: _____ Date Prepared: _____

Procurement Authority:

Describe the project manager's decision authority and limitations, including at least: budget, signature level, contract changes, negotiation, and technical oversight.

Roles and Responsibilities:

Project Manager:	Procurement Department:
1. *Define the responsibilities of the project manager and their team.* 2. 3. 4. 5.	1. *Describe the responsibilities of the procurement or contracting representative and department.* 2. 3. 4. 5.

Standard Procurement Documents:

1. *List any standard procurement forms, documents, policies, or procedures relevant to procurements.*
2.
3.
4.
5.

Contract Type:

Identify the contract type, incentive or award fees, and the criteria for such fees.

Bonding and Insurance Requirements:

Define bonding or insurance requirements that bidders must meet.

Selection Criteria:

Weight	Criteria
Identify selection criteria and their relative weighting. Include information on independent cost estimates if appropriate.	

Procurement Assumptions and Constraints:

Identify and document relevant assumptions and constraints related to the procurement process.

PROCUREMENT MANAGEMENT PLAN

Integration Requirements:

WBS	*Define how the contractor's WBS should integrate with the project WBS.*
Schedule	*Define how the contractor's schedule should integrate with the project schedule, including milestones and long lead items.*
Documentation	*Define any documentation needed from the contractor and how that documentation will integrate with project documentation.*
Risk	*Define how risk identification, analysis, and tracking will integrate with the project risk management.*
Performance Reporting	*Define how the contractor's performance reporting should integrate with the project status reporting, including information on scope, schedule, and cost status reporting.*

3.34 SOURCE SELECTION CRITERIA

The Source Selection Criteria is used to determine and rate the criteria that will be used to evaluate bid proposals. This is a multi-step process.

1. The criteria to evaluate bid responses is determined.
2. A weight is assigned to each criterion. The sum of all the criteria equals 100 percent.
3. The range of ratings for each criterion is determined, such as 1–5 or 1–10.
4. The performance necessary for each rating is defined.
5. Each proposal is evaluated against the criteria and is rated accordingly.
6. The weight is multiplied by the rate and a score for each criterion is derived.
7. The scores are totaled and the highest total score is the winner of the bid.

Evaluation criteria may include:

- Technical expertise
- Prior experience
- Schedule
- Management control and systems
- Price
- Quality
- Intellectual property rights
- Warranty
- Life cycle cost
- Risk management approach

This is just a sample of information that can be evaluated. Use the information on your project to tailor the criteria and rating to best meet your needs.

The Source Selection Criteria is an output from the process 12.1 Plan Procurements in the *PMBOK® Guide—Fourth Edition*.

SOURCE SELECTION CRITERIA

Project Title: _____

Date Prepared: _____

	1	2	3	4	5
Criteria 1					
Criteria 2					
Criteria 3					
Criteria 4					
Criteria 5					

	Weight	Candidate 1 Rating	Candidate 1 Score	Candidate 2 Rating	Candidate 2 Score	Candidate 3 Rating	Candidate 3 Score
Criteria 1							
Criteria 2							
Criteria 3							
Criteria 4							
Criteria 5							
Totals							

Page 1 of 1

SOURCE SELECTION CRITERIA

Project Title: _____   **Date Prepared:** _____

	1	2	3	4	5
Criteria 1	Describe what a 1 means for the criteria. For example, for experience, it may mean that the bidder has no prior experience.	Describe what a 2 means for the criteria. For example, for experience, it may mean that the bidder has done 1 similar job.	Describe what a 3 means for the criteria. For example, for experience, it may mean that the bidder has 3 to 5 similar jobs.	Describe what a 4 means for the criteria. For example, for experience, it may mean that the bidder has done 5 to 10 similar jobs.	Describe what a 5 means for the criteria. For example, for experience, it may mean that the job is the bidder's core competency.
Criteria 2					
Criteria 3					
Criteria 4					
Criteria 5					

	Weight	Candidate 1 Rating	Candidate 1 Score	Candidate 2 Rating	Candidate 2 Score	Candidate 3 Rating	Candidate 3 Score
Criteria 1	Enter the weight for each criteria.	Enter the rating per the above chart.	Weight x rate	Enter the rating per the above chart.	Weight x rate	Enter the rating per the above chart.	Weight x rate
Criteria 2							
Criteria 3							
Criteria 4							
Criteria 5							
Totals	1.0		Sum all scores.		Sum all scores.		Sum all scores.

3.35 PROJECT MANAGEMENT PLAN

The Project Management Plan describes how the team will execute, monitor, control, and close the project. While it has some unique information, it is primarily comprised of all the subsidiary management plans and the baselines. The Project Management Plan combines all this information into a cohesive and integrated approach to managing the project. Typical information includes:

* Selected life cycle
* Variance thresholds
* Baseline management
* Timing and types of reviews
* Tailoring decisions
* Specific approaches to meet project objectives

The Project Management Plan contains plans for managing all the other knowledge areas as well as other specific aspects of the proejct. These take the form of subsidiary management plans and can include:

* Requirements Management Plan
* Scope Management Plan*
* Schedule Management Plan*
* Cost Management Plan*
* Quality Management Plan
* Process Improvement Plan
* Human Resource Management Plan
* Communications Management Plan
* Risk Management Plan
* Procurement Management Plan
* Change Management Plan
* Configuration Management Plan

The Project Management Plan also contains baselines. Common baselines include:

* Scope Baseline
* Schedule Baseline
* Cost Performance Baseline

These can be combined into a Performance Measurement Baseline if earned value measurement techniques are being used

In addition, any other relevant, project-specific information that will be used to manage the project is recorded in the Project Management Plan.

---

*These forms are addressed in the beginning of Chapters 5, 6, and 7 of the *PMBOK® Guide*, respectively. They are generally not developed as a specific plan, however, information on how to address each of these knowledge areas can be entered into the Project Management Plan as appropriate. They are therefore listed as subsidiary management plans here.

The Project Management Plan can receive information from all the subsidiary management plans and baselines.

It provides information to:

- Quality Audits
- Project Performance Reports
- Change Requests
- Variance Analysis
- Earned Value Status
- Product Acceptance
- Contractor Status Report
- Contract Close-out
- Project Close-out
- Lessons Learned

The Project Management Plan is an output from the process 4.1 Develop Project Management Plan in the *PMBOK® Guide*—Fourth Edition.

PROJECT MANAGEMENT PLAN

Project Title: _____   **Date Prepared:** _____

Project Life Cycle:

Variance and Baseline Management:

Schedule Variance Threshold:	Schedule Baseline Management:
Cost Variance Threshold:	Cost Baseline Management:
Scope Variance Threshold:	Scope Baseline Management:
Quality Variance Threshold:	Performance Requirements Management:

Project Reviews:

Tailoring Decisions:

PROJECT MANAGEMENT PLAN

Project-Specific Considerations:

Subsidiary Management Plans:

Area	Approach
Requirements Management Plan	
Scope Management Plan	
Schedule Management Plan	
Cost Management Plan	
Quality Management Plan	
Process Improvement Plan	
Human Resource Management Plan	
Communications Management Plan	
Risk Management Plan	
Procurement Management Plan	
Change Management Plan	
Configuration Management Plan	

Baselines:
Attach all project baselines.

PROJECT MANAGEMENT PLAN

Project Title: _____   Date Prepared: _____

Project Life Cycle:

Describe the life cycle that will be used to accomplish the project. This may include phases, periods within the phases, and deliverables for each phase.

Variances and Baseline Management:

Schedule Variance Threshold: *Define acceptable schedule variances, variances that indicate a warning, and variances that are unacceptable.*	Schedule Baseline Management: *Describe how the schedule baseline will be managed, including responses to acceptable, warning, and unacceptable variances. Define circumstances that would trigger preventive or corrective action and when the change control process would be enacted.*
Cost Variance Threshold: *Define acceptable cost variances, variances that indicate a warning, and variances that are unacceptable.*	Cost Baseline Management: *Describe how the cost performance baseline will be managed, including responses to acceptable, warning, and unacceptable variances. Define circumstances that would trigger preventive or corrective action and when the change control process would be enacted.*
Scope Variance Threshold: *Define acceptable scope variances, variances that indicate a warning, and variances that are unacceptable.*	Scope Baseline Management: *Describe how the scope baseline will be managed, including responses to acceptable, warning, and unacceptable variances. Define circumstances that would trigger preventive or corrective action or defect repair and when the change control process would be enacted.*
Quality Variance Threshold: *Define acceptable performance variances, variances that indicate a warning, and variances that are unacceptable.*	Performance Requirements Management: *Describe how the performance requirements will be managed, including responses to acceptable, warning, and unacceptable variances. Define circumstances that would trigger preventive or corrective action or defect repair and when the change control process would be enacted.*

Project Reviews:

List any project reviews, for example phase gate reviews, customer product reviews, quality reviews, etc.

Tailoring Decisions:

Indicate any decisions made to combine, omit, or expand project management processes. This may include defining the specific processes used in each life cycle phase and the whether it is a summary or detailed application of specific processes.

PROJECT MANAGEMENT PLAN

Project-Specific Considerations:

This may include specific information about the environment, stakeholders, product integration, or any other aspects of the project that warrant special attention.

Subsidiary Management Plans:

Area	Approach
Either define the approach to each subsidiary plan in narrative form or indicate the plan is an attachment.	
Requirements Management Plan	
Scope Management Plan	
Schedule Management Plan	
Cost Management Plan	
Quality Management Plan	
Process Improvement Plan	
Human Resource Management Plan	
Communications Management Plan	
Risk Management Plan	
Procurement Management Plan	
Change Management Plan	
Configuration Management Plan	

Baselines:
Attach all project baselines.

3.36 CONFIGURATION MANAGEMENT PLAN

The Configuration Management Plan is a component of the Project Management Plan. It describes how functional and physical attributes of products will be controlled on the project. It may or may not be aligned with a Change Management Plan. Typical information includes:

- Configuration management approach
- Components subject to configuration control
- Component identification conventions
- Documents subject to configuration and/or version control
- Document identification conventions
- Configuration change control requirements
- Configuration verification and audit procedures
- Configuration management roles and responsibilities

The Configuration Management Plan is related to:

- Change Management Plan

It provides information to:

- Project Management Plan

CONFIGURATION MANAGEMENT PLAN

Project Title: _____ Date Prepared: _____

Configuration Management Approach:

Configuration Identification:

Component	Identification Conventions

Document Configuration Control:

Configuration-Controlled Documents	Identification Conventions
Version-Controlled Documents	Identification Conventions

Configuration Change Control:

Configuration Verification and Audit:

CONFIGURATION MANAGEMENT PLAN

Configuration Management Roles and Responsibilities:

Roles	Responsibilities

Attach any relevant forms used in the configuration control process.

CONFIGURATION MANAGEMENT PLAN

Project Title: _____   Date Prepared: _____

Configuration Management Approach:

Describe the degree of configuration control, the relationship to change control, and how configuration management will integrate with other aspects of project management.

Configuration Identification:

Component	Identification Conventions
Identify the attributes of the product that will be tracked. For example, if the project is to build a new car, define the deliverables that make up the car that will be identified, tracked, and managed. This could include the engine, the transmission, the chassis, and the frame as components that are subject to configuration control. Each of the parts and pieces that make up these deliverables would then be listed and tracked throughout the life of the product.	*Describe how the elements of individual deliverables will be identified. For example, determine naming conventions for all parts of the engine, transmission, chassis, and frame.*

Document Configuration Control:

Configuration-Controlled Documents	Identification Conventions
List all documents that are subject to configuration control.	*Describe how the documents will be named and identified.*
Version-Controlled Documents	**Identification Conventions**
List all the documents that are subject to version control.	*Describe how the documents will named and identified.*

Configuration Change Control:

Describe the process and approvals required to change a configuration item's attributes and to rebaseline them. This includes physical components and documents under configuration control.

Configuration Verification and Audit:

Describe the frequency, timing, and approach to conducting configuration audits to ensure that the physical attributes of the product are consistent with plans, requirements, and configuration documentation.

CONFIGURATION MANAGEMENT PLAN

Configuration Management Roles and Responsibilities:

Roles	Responsibilities
List the roles involved in configuration management.	*List the responsibilities and activities associated with the roles.*

Attach any relevant forms used in the configuration control process.

3.37 CHANGE MANAGEMENT PLAN

The Change Management Plan is a component of the Project Management Plan. It describes how change will be managed on the project. It may or may not be aligned with a Configuration Management Plan. Typical information includes:

- Structure and membership of a change control board
- Definitions of change
- Change control board
  - Roles
  - Responsibilities
  - Authority
- Change management process
  - Change request submittal
  - Change request tracking
  - Change request review
  - Change request disposition

The Change Management Plan is related to:

- Change Log
- Change Request form
- Configuration Management Plan

It provides information to:

- Project Management Plan

CHANGE MANAGEMENT PLAN

Project Title: _____ **Date Prepared:** _____

Change Management Approach:

Definitions of Change:

Schedule change:
Budget change:
Scope change:
Project document changes:

Change Control Board:

Name	Role	Responsibility	Authority

Change Control Process:

Change request submittal	
Change request tracking	
Change request review	
Change request disposition	

Attach relevant forms used in the change control process.

CHANGE MANAGEMENT PLAN

Project Title: _____ **Date Prepared:** _____

Change Management Approach:

Describe the degree of change control, the relationship to configuration management, and how change control will integrate with other aspects of project management.

Definitions of Change:

Schedule change:Define a schedule change versus a schedule revision. Indicate when a schedule variance needs to go through the change control process to be rebaselined.
Budget change:Define a budget change versus a budget update. Indicate when a budget variance needs to go through the change control process to be rebaselined.
Scope change:Define a scope change versus a change in approach. Indicate when a scope variance needs to go through the change control process to be rebaselined.
Project document changes:Define when updates to project management documents or other project documents need to go through the change control process to be rebaselined.

Change Control Board:

Name	Role	Responsibility	Authority
Individual's name.	*Position on the change control board.*	*Responsibilities and activities required of the role.*	*Authority level for approving or rejecting changes.*

Change Control Process:

Change request submittal	*Describe the process used to submit change requests, including who receives requests and any special forms, policies, or procedures that need to be used.*
Change request tracking	*Describe the process for tracking change requests from submittal to final disposition.*
Change request review	*Describe the process used to review change requests, including analysis of impact on project objectives such as schedule, scope, cost, etc.*
Change request disposition	*Describe the possible outcomes, such as accept, defer, reject.*

Attach relevant forms used in the change control process.

Executing Forms

4.1 EXECUTING PROCESS GROUP

The purpose of the Executing Process Group is to carry out the work necessary to meet the project objectives. There are eight processes in the Executing Process Group.

- Direct and Manage Project Execution
- Perform Quality Assurance
- Acquire Project Team
- Develop Project Team
- Manage Project Team
- Distribute Information
- Manage Stakeholder Expectations
- Conduct Procurements

The intent of the Executing Process Group is to at least:

- Create the deliverables
- Manage project quality
- Manage the project team
- Carry out project communications
- Report progress
- Manage changes
- Manage stakeholders
- Bid and award contracts

In these processes, the main work of the project is carried out and the majority of the funds are expended. To be effective, the project manager must coordinate project resources, manage changes, report progress, and manage stakeholders while completing the project deliverables.

The forms used to document project execution include:

- Team Member Status Report
- Change Request
- Change Log
- Decision Log
- Quality Audit
- Team Directory
- Team Operating Agreement
- Team Performance Assessment
- Team Member Performance Appraisal
- Issue Log

4.2 TEAM MEMBER STATUS REPORT

The Team Member Status Report is filled out by team members and submitted to the project manager on a regular basis. It tracks schedule and cost status for the current reporting period and provides planned information for the next reporting period. Status reports also identify new risks and issues that have arisen in the current reporting period. Typical information includes:

- Activities planned for the current reporting period
- Activities accomplished in the current reporting period
- Activities planned but not accomplished in the current reporting period
- Root causes of variances
- Funds spent in the current reporting period
- Funds planned to be spent for the current reporting period
- Root causes of variances
- Quality variances identified in the current reporting period
- Planned corrective or preventive action
- Activities planned for the next reporting period
- Costs planned for the next reporting period
- New risks identified
- Issues
- Comments

This information is generally compiled by the project manager into a Project Performance Report.

TEAM MEMBER STATUS REPORT

Project Title: _____   Date Prepared: _____

Team Member: _____   Role: _____

Activities Planned for This Reporting Period:

1. 2. 3. 4. 5. 6.

Activities Accomplished This Reporting Period:

1. 2. 3. 4. 5. 6.

Activities Planned but Not Accomplished This Reporting Period:

1. 2. 3. 4.

Root Cause of Variances:

Funds Spent This Reporting Period:

Funds Planned to Be Spent This Reporting Period:

Root Cause of Variances:

TEAM MEMBER STATUS REPORT

Quality Variances Identified This Period:

Planned Corrective or Preventive Action:

Activities Planned for Next Reporting Period:

1.
2.
3.
4.
5.
6.

Costs Planned for Next Reporting Period:

New Risks Identified:

Issues:

Comments:

TEAM MEMBER STATUS REPORT

Project Title: _____   Date Prepared: _____

Team Member: _____   Role: _____

Activities Planned for This Reporting Period:

1. *List all activities scheduled for this period, including work to be started, continued, or completed.*
2.
3.
4.
5.
6.

Activities Accomplished This Reporting Period:

1. *List all activities accomplished this period, including work that was started, continued, or completed.*
2.
3.
4.
5.
6.

Activities Planned but Not Accomplished This Reporting Period:

1. *List all activities that were scheduled for this period, but not started, continued, or completed.*
2.
3.
4.

Root Cause of Variances:

For any work that was not accomplished as scheduled, identify the cause of the variance.

Funds Spent This Reporting Period:

Record funds spent this period.

Funds Planned to Be Spent This Reporting Period:

Record the funds that were planned to be spent for this period.

Root Cause of Variances:

For any expenditures that were over or under plan, identify the cause of the variance. Include information on labor variance versus material variances.

TEAM MEMBER STATUS REPORT

Quality Variances Identified This Period:

Identify any product performance or quality variances.

Planned Corrective or Preventive Action:

Identify any actions needed to recover cost, schedule, or quality variances or prevent future variances.

Activities Planned for Next Reporting Period:

1. *List all activities scheduled for next period, including work to be started, continued, or completed.*
2.
3.
4.
5.
6.

Costs Planned for Next Reporting Period:

Identify funds planned to be expended next period.

New Risks Identified:

Identify any new risks that have arisen this period. These risks should be recorded in the Risk Register as well.

Issues:

Identify any new issues that have arisen this period. These issues should be recorded in the Issue Log as well.

Comments:

Record any comments that add relevance to the report.

4.3 CHANGE REQUEST

A Change Request is used to change any aspect of the project. It can pertain to project, product, documents, requirements, or any other aspect of the project. Upon completion, it is submitted to the change control board or other similar body for review. Typical information includes:

- Person requesting the change
- An identifier, such as the change number
- Category of change
- Detailed description of the proposed change
- Justification for the proposed change
- Impacts of the proposed change
  - Scope
  - Quality
  - Requirements
  - Cost
  - Schedule
  - Project documents
- Disposition of change
- Justification
- Signatures of change control board

The change request form can result from these processes:

- Direct and Manage Project Execution
- Verify Scope
- Control Schedule
- Perform Quality Assurance
- Manage Project Team
- Report Performance
- Plan Procurements
- Administer Procurements

- Monitor and Control Project Work
- Control Scope
- Control Costs
- Perform Quality Control
- Manage Stakeholder Expectations
- Monitor and Control Risks
- Conduct Procurements

The Change Request form is related to:

- Change Log
- Change Management Plan

It provides information to the following process:

- Perform Integrated Change Control

CHANGE REQUEST

Project Title: _____   Date Prepared: _____

Person Requesting Change: _____   Change Number: _____

Category of Change:

☐ Scope   ☐ Quality   ☐ Requirements

☐ Cost   ☐ Schedule   ☐ Documents

Detailed Description of Proposed Change:

Justification for Proposed Change:

Impacts of Change:

Scope	☐ Increase	☐ Decrease	☐ Modify
Description:			
Quality	☐ Increase	☐ Decrease	☐ Modify
Description:			

CHANGE REQUEST

Requirements	☐ Increase	☐ Decrease	☐ Modify

Description:

Cost	☐ Increase	☐ Decrease	☐ Modify

Description:

Schedule	☐ Increase	☐ Decrease	☐ Modify

Description:

Project Documents

Comments:

CHANGE REQUEST

Disposition   ☐ Approve   ☐ Defer   ☐ Reject

Justification:

Change Control Board Signatures:

Name	Role	Signature

Date: _____

CHANGE REQUEST

Project Title: _____   **Date Prepared:** _____

Person Requesting Change: _____   **Change Number:** _____

Category of Change *(Check a box to indicate the category of change.):*

☐ Scope   ☐ Quality   ☐ Requirements

☐ Cost   ☐ Schedule   ☐ Documents

Detailed Description of Proposed Change:

Describe change proposed.

Justification for Proposed Change:

Indicate the reason for the change.

Impacts of Change:

Scope	☐ Increase	☐ Decrease	☐ Modify
Description: *Describe the impact of the proposed change on the project or product scope.*			
Quality	☐ Increase	☐ Decrease	☐ Modify
Description: *Describe the impact of the proposed change on the project or product quality.*			

CHANGE REQUEST

Requirements	☐ Increase	☐ Decrease	☐ Modify

Description:

Describe the impact of the proposed change on the project or product requirements.

Cost	☐ Increase	☐ Decrease	☐ Modify

Description:

Describe the impact of the proposed change on the project budget or cost estimates.

Schedule	☐ Increase	☐ Decrease	☐ Modify

Description:

Describe the impact of the proposed change on the schedule and whether it will cause a delay on the critical path.

Project Documents

Describe changes needed to project documents.

Comments:

Any comments that will clarify information on the change request.

CHANGE REQUEST

Disposition   ☐ Approve   ☐ Defer   ☐ Reject

Justification:

Justification for the change request disposition.

Change Control Board Signatures:

Name	Role	Signature

Date: _____

4.4 CHANGE LOG

The Change Log is a dynamic document that is kept througout the project. It is used to track changes from request through final disposition. Typical information includes:

- Change ID
- Category
- Description of change
- Submitter
- Submission date
- Status
- Disposition

The change log is related to:

- Change Request
- Change Management Plan

CHANGE LOG

Project Title: _____

Date Prepared: _____

Change ID	Category	Description of Change	Submitted by	Submission Date	Status	Disposition

CHANGE LOG

Project Title: _____ **Date Prepared:** _____

Change ID	Category	Description of Change	Submitted by	Submission Date	Status	Disposition
Identifier.	_From the change request form._	_Description of the proposed change._	_Person requesting the change._	_Date change was submitted._	_Open, closed, pending, etc._	_Approved, deferred, or rejected._

4.5 DECISION LOG

The Decision Log is a dynamic document that is kept throughout the project. Frequently there are alternatives in developing a product or managing a project. Using a Decision Log can help keep track of the decisions that were made, who made them, and when they were made. A Decision Log can include:

- Identifier
- Category
- Decision
- Responsible party
- Date
- Comments

Use the information from your project to tailor the form to best meet your needs.

DECISION LOG

Project Title: _____

Date Prepared: _____

ID	Category	Decision	Responsible Party	Date	Comments

DECISION LOG

Project Title: _____   Date Prepared: _____

ID	Category	Decision	Responsible Party	Date	Comments
Identifier.	*Type of decision: technical, project, process, etc.*	*Description of the decision.*	*Person authorized and making the decision.*	*Date of decision.*	*Any further information to clarify the alternatives considered, reasons the decision was made, and impacts of the decision.*

4.6 QUALITY AUDIT

A Quality Audit is a technique that employs a structured, independent review to project and/or product elements. Any aspect of the project or product can be audited. Common areas for audit include:

- Project processes
- Project documents
- Product requirements
- Product documents
- Implementation of approved changes
- Implementation of corrective or preventive action
- Defect or deficiency repair
- Compliance with organizational policies and procedures
- Compliance with the quality plan

Additional audit information can include:

- Good practices to share
- Areas for improvement
- Description of deficiencies or defects

Defects or deficiencies should include action items, a responsible party, and be assigned a due date for compliance.

Audits should be tailored to best meet the needs of the project.

Results from the audit may necessitate a Change Request, including preventive or corrective action, and defect repair.

A Quality Audit is a technique from the process 8.2 Perform Quality Assurance in the *PMBOK*® Guide—Fourth Edition.

QUALITY AUDIT

Project Title: _____   Date Prepared: _____

Project Auditor: _____   Audit Date: _____

Area Audited:

☐ Project	☐ Project processes	☐ Project documents
☐ Product	☐ Product requirements	☐ Product documents
☐ Approved change implementation	☐ Corrective or preventive action implementation	☐ Defect/deficiency repair
☐ Quality Management Plan	☐ Organizational policies	☐ Organizational procedures

Description of Good Practices to Share:

Description of Areas for Improvement:

Description of Deficiencies or Defects:

ID	Defect	Action	Responsible Party	Due Date

Comments:

QUALITY AUDIT

Project Title: _____   **Date Prepared:** _____

Project Auditor: _____   **Audit Date:** _____

Area Audited *(Indicate what was audited):*

☐ Project	☐ Project processes	☐ Project documents
☐ Product	☐ Product requirements	☐ Product documents
☐ Approved change implementation	☐ Corrective or preventive action implementation	☐ Defect/deficiency repair
☐ Quality Management Plan	☐ Organizational policies	☐ Organizational procedures

Description of Good Practices to Share:

Describe any good or best practices that can be shared with other projects.

Description of Areas for Improvement:

Describe any areas that need improvement and the specific improvements or measurements that need to be achieved.

Description of Deficiencies or Defects:

ID	Defect	Action	Responsible Party	Due Date
	Describe deficiencies or defects.	*Describe the action needed to be taken to correct defects.*	*Assigned person*	*Due date*

Comments:

4.7 TEAM DIRECTORY

The Team Directory lists the project team members and their primary contact information. It is particularly useful on virtual teams when team members often have not met one another and may work in different time zones. The contents of the team directory include:

- Name
- Role
- Department
- E-mail address
- Mobile and work phone numbers
- Work hours

Additional information can include the geographical location and organization that the individual works for. Use the information from your project to tailor the form to best meet your needs.

The Team Directory is compiled when team members are assigned through the Acquire Team Members process. It provides information to the Develop Project Team and Manage Project Team processes.

TEAM DIRECTORY

Project Title: _____   Date Prepared: _____

Name	Role	Department	E-mail	Phone Numbers (Mobile and Work)	Work Hours

TEAM DIRECTORY

Project Title: _____   Date Prepared: _____

Name	Role	Department	E-mail	Phones Numbers (Mobile and Work)	Work Hours
Name the person likes to be called.	_Role on the team._	_Functional department._	_E-mail address._	_Phone numbers._	_Work hours or time zones._

4.8 TEAM OPERATING AGREEMENT

The Team Operating Agreement is used to establish ground rules and guidelines for the team. It is particularly useful on virtual teams and teams that are comprised of members from different organizations. Using a Team Operating Agreement can help establish expectations and agreements on working effectively together. The contents of the Team Operating Agreement include:

- Team values and principles
- Meeting guidelines
- Communication guidelines
- Decision-making process
- Conflict management approach

Additional information can be included as appropriate for the individual project and the individual team members. Use the information on your project to tailor the form to best meet your needs.

The Team Operating Agreement is developed when team members are assigned through the Acquire Team Members process.

TEAM OPERATING AGREEMENT

Project Title: _____   Date Prepared: _____

Team Values and Principles:

Meeting Guidelines:

Communication Guidelines:

Decision-Making Process:

Conflict Management Approach:

TEAM OPERATING AGREEMENT

Other Agreements:

Signature:

_____

_____

_____

_____

_____

_____

_____

Date:

_____

_____

_____

_____

_____

_____

TEAM OPERATING AGREEMENT

Project Title: _____   Date Prepared: _____

Team Values and Principles:

List values and principles that the team agrees to operate within.
Examples include mutual respect, operating from fact not opinion, etc.

Meeting Guidelines:

Identify guidelines that will keep meetings productive.
Examples include decision makers must be present, start on time, stick to the agenda, etc.

Communication Guidelines:

List guidelines used for effective communication.
Examples include everyone voices their opinion, no dominating the conversation, no interrupting, not using inflammatory language, etc.

Decision-Making Process:

Describe the process used to make decisions. Indicate the relative power of the project manager for decision making as well as any voting procedures. Also indicate the circumstances under which a decision can be revisited.

Conflict Management Approach:

Describe the approach to managing conflict, when a conflict will be escalated, when it should be tabled for later discussion, etc.

TEAM OPERATING AGREEMENT

Other Agreements:

List any other agreements or approaches to ensuring a collaborative and productive working relationship among team members.

Signature:

Date:

4.9 TEAM PERFORMANCE ASSESSMENT

The Team Performance Assessment is used to review technical performance and interpersonal competencies of the team as a whole. It also addresses team morale and areas for team performance improvement. The contents of the Team Performance Assessment can include:

- Technical performance
  - Scope
  - Quality
  - Schedule
  - Cost
- Interpersonal competency
  - Communication
  - Collaboration
  - Conflict management
  - Decision making
- Team morale
- Areas for development

This is just a sample of information that can be evaluated. Use the information from your project to tailor the form to best meet your needs.

The Team Performance Assessment is an output from the process 9.3 Develop Project Team in the *PMBOK® Guide*—Fourth Edition.

It provides information to the Manage Project Team process.

TEAM PERFORMANCE ASSESSMENT

Project Title: _____   Date Prepared: _____

Technical Performance:

Scope	☐ Exceeds Expectations	☐ Meets Expectations	☐ Needs Improvement
Comments:			

Quality	☐ Exceeds Expectations	☐ Meets Expectations	☐ Needs Improvement
Comments:			

Schedule	☐ Exceeds Expectations	☐ Meets Expectations	☐ Needs Improvement
Comments:			

Cost	☐ Exceeds Expectations	☐ Meets Expectations	☐ Needs Improvement
Comments:			

Interpersonal Competency:

Communication	☐ Exceeds Expectations	☐ Meets Expectations	☐ Needs Improvement
Comments:			

Collaboration	☐ Exceeds Expectations	☐ Meets Expectations	☐ Needs Improvement
Comments:			

Conflict Management	☐ Exceeds Expectations	☐ Meets Expectations	☐ Needs Improvement
Comments:			

Decision Making	☐ Exceeds Expectations	☐ Meets Expectations	☐ Needs Improvement
Comments:			

TEAM PERFORMANCE ASSESSMENT

Team Morale:

Comments:

Areas for Development:

Area	Approach	Actions

TEAM PERFORMANCE ASSESSMENT

Project Title: _____ Date Prepared: _____

Technical Performance:

Scope	☐ Exceeds Expectations	☐ Meets Expectations	☐ Needs Improvement
Comments: *Include comments that describe instances or aspects of scope performance that explain the rating.*			
Quality	☐ Exceeds Expectations	☐ Meets Expectations	☐ Needs Improvement
Comments: *Include comments that describe instances or aspects of quality performance that explain the rating.*			
Schedule	☐ Exceeds Expectations	☐ Meets Expectations	☐ Needs Improvement
Comments: *Include comments that describe instances or aspects of schedule performance that explain the rating.*			
Cost	☐ Exceeds Expectations	☐ Meets Expectations	☐ Needs Improvement
Comments: *Include comments that describe instances or aspects of cost performance that explain the rating.*			

Interpersonal Competency:

Communication	☐ Exceeds Expectations	☐ Meets Expectations	☐ Needs Improvement
Comments: *Include comments that describe instances or aspects of communication that explain the rating.*			
Collaboration	☐ Exceeds Expectations	☐ Meets Expectations	☐ Needs Improvement
Comments: *Include comments that describe instances or aspects of collaboration that explain the rating.*			
Conflict Management	☐ Exceeds Expectations	☐ Meets Expectations	☐ Needs Improvement
Comments: *Include comments that describe instances or aspects of conflict management that explain the rating.*			
Decision Making	☐ Exceeds Expectations	☐ Meets Expectations	☐ Needs Improvement
Comments: *Include comments that describe instances or aspects of decision making that explain the rating.*			

TEAM PERFORMANCE ASSESSMENT

Team Morale:

Comments:
Describe the overall team morale.

Areas for Development:

Area	Approach	Actions
List technical or interpersonal areas for development.	*Describe the development approach, such as mentoring, training, etc.*	*List the actions necessary to implement the development approach.*

4.10 TEAM MEMBER PERFORMANCE ASSESSMENT

The Team Member Performance Assessment is used to review technical performance, interpersonal competencies, and strengths and weaknesses of individual team members. On many projects, the project manager does not provide formal team member assessment or evaluation. The performance assessment can be done very informally depending on the organization's culture. The contents of the Team Member Performance Assessment can include:

- Technical performance
  - Scope
  - Quality
  - Schedule
  - Cost
- Interpersonal competency
  - Communication
  - Collaboration
  - Conflict management
  - Decision making
  - Leadership
- Strengths and weaknesses
- Areas for development

This is just a sample of information that can be evaluated. Use the information from your project to tailor the form to best meet your needs.

TEAM MEMBER PERFORMANCE ASSESSMENT

Project Title: _____   Date Prepared: _____

Technical Performance:

Scope	☐ Exceeds Expectations	☐ Meets Expectations	☐ Needs Improvement
Comments:			
Quality	☐ Exceeds Expectations	☐ Meets Expectations	☐ Needs Improvement
Comments:			
Schedule	☐ Exceeds Expectations	☐ Meets Expectations	☐ Needs Improvement
Comments:			
Cost	☐ Exceeds Expectations	☐ Meets Expectations	☐ Needs Improvement
Comments:			

Interpersonal Competency:

Communication	☐ Exceeds Expectations	☐ Meets Expectations	☐ Needs Improvement
Comments:			
Collaboration	☐ Exceeds Expectations	☐ Meets Expectations	☐ Needs Improvement
Comments:			
Conflict Management	☐ Exceeds Expectations	☐ Meets Expectations	☐ Needs Improvement
Comments:			
Decision Making	☐ Exceeds Expectations	☐ Meets Expectations	☐ Needs Improvement
Comments:			
Leadership	☐ Exceeds Expectations	☐ Meets Expectations	☐ Needs Improvement
Comments:			

TEAM MEMBER PERFORMANCE ASSESSMENT

Strengths:

Weaknesses:

Areas for Development:

Area	Approach	Actions

Additional Comments:

TEAM MEMBER PERFORMANCE ASSESSMENT

Project Title: _____ Date Prepared: _____

Technical Performance:

Scope	☐ Exceeds Expectations	☐ Meets Expectations	☐ Needs Improvement
Comments: *Include comments that describe instances or aspects of scope performance that explain the rating.*			
Quality	☐ Exceeds Expectations	☐ Meets Expectations	☐ Needs Improvement
Comments: *Include comments that describe instances or aspects of quality performance that explain the rating.*			
Schedule	☐ Exceeds Expectations	☐ Meets Expectations	☐ Needs Improvement
Comments: *Include comments that describe instances or aspects of schedule performance that explain the rating.*			
Cost	☐ Exceeds Expectations	☐ Meets Expectations	☐ Needs Improvement
Comments: *Include comments that describe instances or aspects of cost performance that explain the rating.*			

Interpersonal Competency:

Communication	☐ Exceeds Expectations	☐ Meets Expectations	☐ Needs Improvement
Comments: *Include comments that describe instances or aspects of communication that explain the rating.*			
Collaboration	☐ Exceeds Expectations	☐ Meets Expectations	☐ Needs Improvement
Comments: *Include comments that describe instances or aspects of collaboration that explain the rating.*			
Conflict Management	☐ Exceeds Expectations	☐ Meets Expectations	☐ Needs Improvement
Comments: *Include comments that describe instances or aspects of conflict management that explain the rating.*			
Decision Making	☐ Exceeds Expectations	☐ Meets Expectations	☐ Needs Improvement
Comments: *Include comments that describe instances or aspects of decision making that explain the rating.*			
Leadership	☐ Exceeds Expectations	☐ Meets Expectations	☐ Needs Improvement
Comments: *Include comments that describe instances or aspects of leadership that explain the rating.*			

TEAM MEMBER PERFORMANCE ASSESSMENT

Strengths:

Describe individual technical and interpersonal strengths that the team member possesses. Give explicit examples.

Weaknesses:

Describe individual technical and interpersonal weaknesses that the team member possesses. Give explicit examples.

Areas for Development:

Area	Type	Actions
List technical or interpersonal areas for development.	*Describe the development approach, such as mentoring, training, etc.*	*List the actions necessary to implement the development approach.*

Additional Comments:

Any comments that provide additional insight or information into the team member's performance.

4.11 ISSUE LOG

The Issue Log is a dynamic document that is kept throughout the project. An issue is a point or matter in question or in dispute, one that is not settled and is under discussion, or one over which there are opposing views or disagreements. An issue can also be a risk event that has occurred and must now be dealt with. An Issue Log includes:

- Identifier
- Category
- Issue
- Impact on objectives
- Responsible party
- Actions
- Status
- Due date
- Comments

Additional information can include the source of the issue and the urgency. Use the information from your project to tailor the form to best meet your needs.

The Issue Log is a tool used in the process 9.4 Manage Project Team and an input in the process 10.4 Manage Stakeholder Expectations in the *PMBOK*® Guide—Fourth Edition.

ISSUE LOG

Project Title: _____

Date Prepared: _____

Issue ID	Category	Issue	Impact on Objectives	Urgency

Responsible Party	Actions	Status	Due Date	Comments

ISSUE LOG

Project Title: _____ Date Prepared: _____

Issue ID	Category	Issue	Impact on Objectives	Urgency
Identifier.	Type of issue; stakeholder, decision, etc.	Define the issue.	Describe the degree of impact on various project objectives.	High, medium, or low.

Responsible Party	Actions	Status	Due Date	Comments
Person assigned to follow up.	Actions needed to address and resolve the issue.	Open or closed.	Date issue needs to be resolved.	Any clarifying comments about the issue, the resolution, or other fields on the form.

Monitoring and Control Forms

5.1 MONITORING AND CONTROLLING PROCESS GROUP

The purpose of the Monitoring and Controlling Process Group is to review project work results and compare them to planned results. A significant variance indicates the need for preventive actions, corrective actions, or change requests. There are 10 processes in the Monitoring and Controlling Process Group:

- Monitor and Control Project Work
- Perform Integrated Change Control
- Verify Scope
- Control Scope
- Control Schedule

- Control Costs
- Perform Quality Control
- Report Performance
- Monitor and Control Risks
- Administer Procurements

The intent of the Monitoring and Controlling Process Group is to at least:

- Review and analyze project performance
- Recommend changes and corrective and preventive actions
- Process change requests
- Report project performance
- Respond to risk events
- Manage contractors

Monitoring and controlling takes place throughout the project, from inception to closing. All variances are identified, and all change requests are processed here. The product deliverables are also accepted in the Monitoring and Controlling processes.

The forms used to document these activities include:

- Project Performance Report
- Variance Analysis
- Earned Value Status
- Risk Audit
- Contractor Status Report
- Product Acceptance

5.2 PROJECT PERFORMANCE REPORT

The Project Performance Report is filled out by the project manager and submitted on a regular basis to the sponsor, project portfolio management group, Project Management Office or other project oversight person or group. The information is compiled from the Team Member Status Reports and also includes overall project performance. It contains summary-level information, such as accomplishments, rather than detailed activity-level information. The Project Performance Report tracks schedule and cost status for the current reporting period and provides planned information for the next reporting period. It indicates impacts to milestones and cost reserves as well as indentifying new risks and issues that have arisen in the current reporting period. Typical information includes:

- Accomplishments for the current reporting period
- Accomplishments planned but not completed in the current reporting period
- Root causes of variances
- Impact to upcoming milestones or project due date
- Planned corrective or preventive action
- Funds spent in the current reporting period
- Root causes of variances
- Impact to overall budget or contingency funds
- Planned corrective or preventive action
- Accomplishments planned for the next reporting period
- Costs planned for the next reporting period
- New risks identified
- Issues
- Comments

The Performance Report is an output used in the process 10.5 Report Performance in the *PMBOK® Guide*—Fourth Edition.

PROJECT PERFORMANCE REPORT

Project Title: _____   Date Prepared: _____

Project Manager: _____   Sponsor: _____

Accomplishments for This Reporting Period:

1.
2.
3.
4.
5.
6.

Accomplishments Planned but Not Completed This Reporting Period:

1.
2.
3.
4.

Root Cause of Variances:

Impact to Upcoming Milestones or Project Due Date:

Planned Corrective or Preventive Action:

Funds Spent This Reporting Period:

Root Cause of Variances:

PROJECT PERFORMANCE REPORT

Impact to Overall Budget or Contingency Funds:

Planned Corrective or Preventive Action:

Accomplishments Planned for Next Reporting Period:

1.
2.
3.
4.

Costs Planned for Next Reporting Period:

New Risks Identified:

Issues:

Comments:

PROJECT PERFORMANCE REPORT

Project Title: _____   Date Prepared: _____

Project Manager: _____   Sponsor: _____

Accomplishments for This Reporting Period:

1. *List all work packages or other accomplishments scheduled for completion this period.*
2.
3.
4.
5.

Accomplishments Planned but Not Completed This Reporting Period:

1. *List all work packages or other accomplishments scheduled for this period but not completed.*
2.
3.
4.

Root Cause of Variances:

For any work that was not accomplished as scheduled, identify cause of the variance.

Impact to Upcoming Milestones or Project Due Date:

For any work that was not accomplished as scheduled, identify any impact to upcoming milestones or overall project schedule. Identify any work currently behind on the critical path or if the critical path has changed based on the variance.

Planned Corrective or Preventive Action:

Identify any actions needed to make up schedule variances or prevent future schedule variances.

Funds Spent This Reporting Period:

Record funds spent this period.

Root Cause of Variances:

For any expenditures that were over or under plan, identify cause of the variance. Include information on labor variance versus material variances.

PROJECT PERFORMANCE REPORT

Impact to Overall Budget or Contingency Funds:

For cost variances, indicate impact to the overall project budget or whether contingency funds must be expended.

Planned Corrective or Preventive Action:

Identify any actions needed to recover cost variances or prevent future schedule variances.

Accomplishments Planned for Next Reporting Period:

1. *List all work packages or accomplishments scheduled for completion next period.*
2.
3.
4.

Costs Planned for Next Reporting Period:

Identify funds planned to be expended next period.

New Risks Identified:

Identify any new risks that have arisen this period. These risks should be recorded in the Risk Register as well.

Issues:

Identify any new issues that have arisen this period. These issues should be recorded in the Issue Log as well.

Comments:

Record any comments that add relevance to the report.

5.3 VARIANCE ANALYSIS

Variance Analysis reports collect and assemble information on project performance variance. Common topics are schedule, cost, and quality variances. Information on a Variance Analysis includes:

- Schedule variance
  - Planned results
  - Actual results
  - Variance
  - Root cause
  - Planned response
- Cost variance
  - Planned results
  - Actual results
  - Variance
  - Root cause
  - Planned response
- Quality variance
  - Planned results
  - Actual results
  - Variance
  - Root cause
  - Planned response

Scope variance can be included but is generally indicated by a schedule variance, as more or less scope will have been accomplished over time. The Variance Analysis can be done at a work package, control account, or project level. It can be used to report status to the project manager to the sponsor or for a vendor. Use the information from your project to tailor the form to best meet your needs.

A Variance Analysis is used as a technique in these processes in the *PMBOK® Guide*—Fourth Edition.

- 5.5 Control Scope
- 6.6 Control Schedule
- 7.3 Cost Control Cost
- 10.5 Report Performance

Information in a Variance Analysis may lead to the project manager submitting a Change Request.

VARIANCE ANALYSIS

Project Title: _____   Date Prepared: _____

Schedule Variance:

Planned Result	Actual Result	Variance

Root Cause:

Planned Response:

Cost Variance:

Planned Result	Actual Result	Variance

Root Cause:

Planned Response:

Quality Variance:

Planned Result	Actual Result	Variance

Root Cause:

Planned Response:

VARIANCE ANALYSIS

Project Title: _____   Date Prepared: _____

Schedule Variance:

Planned Result	Actual Result	Variance
Identify the work planned to be accomplished.	*Identify the work actually accomplished.*	*Identify the variance.*

Root Cause:
Describe the root cause of the variance.

Planned Response:
Describe the planned corrective action.

Cost Variance:

Planned Result	Actual Result	Variance
Record the planned costs for the work planned to be accomplished.	*Identify the actual costs for the work accomplished.*	*Identify the variance.*

Root Cause:
Describe the root cause of the variance.

Planned Response:
Describe the planned corrective action.

Quality Variance:

Planned Result	Actual Result	Variance
Describe the planned performance or quality measurements.	*Describe the actual performance or quality measurements.*	*Identify the variance.*

Root Cause:
Describe the root cause of the variance.

Planned Response:
Describe the planned corrective action.

5.4 EARNED VALUE STATUS

An Earned Value Status report shows specific mathematical metrics that are designed to reflect the health of the project by integrating scope, schedule, and cost information. Information can be reported for the current reporting period and on a cumulative basis. Earned Value Status reports can also be used to forecast the total cost of the project. Information that is generally collected includes:

- Budget at completion (BAC)
- Planned value (PV)
- Earned value (EV)
- Actual cost (AC)
- Schedule variance (SV)
- Cost variance (CV)
- Schedule performance index (SPI)
- Cost performance index (CPI)
- Percent planned
- Percent earned
- Percent spent
- Estimates at completion (EAC)
- To complete performance index (TCPI)

Many different equations can be used to calculate the EAC. Two options are presented on this form. Similarly, there are various options to calculate a TCPI. Use the information from your project to determine the best approach for reporting. Information should reflect the most accurate historical data and assumptions for forecasts, and predictions should be documented and justified. Where appropriate, show the equations used to derive estimates.

Earned Value Status Reports can be used as a technique in these processes in the *PMBOK® Guide*—Fourth Edition.

- 6.6 Control Schedule
- 7.3 Control Costs
- 10.5 Report Performance
- 11.6 Monitor and Control Risks
- 12.3 Administer Procurements

EARNED VALUE STATUS REPORT

Project Title: _____   Date Prepared: _____

Budget at Completion (BAC): _____   Overall Status: _____

	Current Reporting Period	Current Period Cumulative	Past Period Cumulative
Planned value (PV)			
Earned value (EV)			
Actual cost (AC)			
Schedule variance (SV)			
Cost variance (CV)			
Schedule performance index (SPI)			
Cost performance index (CPI)			

Root Cause of Schedule Variance:

Impact on Deliverables, Milestones, or Critical Path:

Root Cause of Cost Variance:

Impact on Budget, Contingency Funds, or Reserve:

Percent planned			
Percent earned			
Percent spent			

Estimates at Completion (EAC):

EAC w/CPI [BAC/CPI]			
EAC w/ CPI x SPI [AC+((BAC-EV)/ (CPI x SPI))]			
Selected EAC, Justification, and Explanation			
To complete performance index (TCPI)			

EARNED VALUE STATUS REPORT

Project Title: _____   Date Prepared: _____

Budget at Completion (BAC): _____   Overall Status: _____

	Current Reporting Period	Current Period Cumulative	Past Period Cumulative
Planned value (PV)	Value of the work planned to be accomplished		
Earned value (EV)	Value of the work actually accomplished		
Actual cost (AC)	Cost for the work accomplished		
Schedule variance (SV)	EV-PV		
Cost variance (CV)	EV-AC		
Schedule performance index (SPI)	EV/PV		
Cost performance index (CPI)	EV/AC		
Root Cause of Schedule Variance:			
Describe the cause of any schedule variances.			
Impact on Deliverables, Milestones, or Critical Path:			
Describe the impact on deliverables, milestones, and the critical path and any intended actions to address the variances.			
Root Cause of Cost Variance:			
Describe the cause of any cost variances.			
Impact on Budget, Contingency Funds, or Reserve:			
Describe the impact on the project budget, contingency funds and reserves, and any intended actions to address the variances.			
Percent planned		PV/BAC	
Percent earned		EV/BAC	
Percent spent		AC/BAC	

Estimates at Completion (EAC):

EAC w/CPI [BAC/CPI]			
EAC w/ CPIxSPI [AC+((BAC-EV)/(CPIxSPI))]			
Selected EAC, Justification and explanation			
There are many valid methods of deriving estimates at completion. Two of them are listed above. Whichever method you choose, document the method and justify the approach.			
To complete performance index (TCPI)		(BAC-EV)/(BAC-AC)*	

*Another common equation for TCPI is (BAC-EV)/(EAC-AC).

5.5 RISK AUDIT

Risk Audits are used to evaluate the effectiveness of the risk identification, risk responses, and risk management process as a whole. Information reviewed in a Risk Audit can include:

- Risk event audits
  - Risk events
  - Causes
  - Responses
- Risk response audits
  - Risk event
  - Responses
  - Success
  - Actions for improvement
- Risk management process
  - Process
  - Compliance
  - Tools and techniques used
- Good practices
- Areas for improvement

Results from the audit may necessitate a Change Request including preventive or corrective action.

The Risk Audit is a tool used in the process 11.6 Monitor and Control Risks in the *PMBOK® Guide—*Fourth Edition.

RISK AUDIT

Project Title: _____   Date Prepared: _____

Project Audit: _____   Audit Date: _____

Risk Event Audit:

Event	Cause	Response	Comment

Risk Response Audit:

Event	Response	Successful	Actions to Improve

Risk Management Process Audit:

Process	Followed	Tools and Techniques Used
Plan Risk Management		
Identify Risks		
Perform Qualitative Risk Analysis		
Perform Quantitative Risk Analysis		
Plan Risk Responses		
Monitor and Control Risks		

Description of Good Practices to Share:

Description of Areas for Improvement:

RISK AUDIT

Project Title: _____   Date Prepared: _____

Project Audit: _____   Audit Date: _____

Risk Event Audit:

Event	Cause	Response	Comment
List the event from the risk register.	*Identify the root cause of the event.*	*Describe the response implemented.*	*Discuss if there was any way to have foreseen the event and respond to it more effectively.*

Risk Response Audit:

Event	Response	Successful	Actions to Improve
List the event from the risk register.	*List the risk response.*	*Indicate if the response was successful.*	*Identify any opportunities for improvement in risk response.*

Risk Management Process Audit:

Process	Followed	Tools and Techniques Used
Plan Risk Management	*Indicate if the various processes were followed as indicated in the risk management plan.*	*Identify tools and techniques used in the various risk management processes and whether they were successful.*
Identify Risks		
Perform Qualitative Risk Analysis		
Perform Quantitative Risk Analysis		
Plan Risk Responses		
Monitor and Control Risks		

Description of Good Practices to Share:

Describe any practices that should be shared for use on other projects. Include any recommendations to update and improve risk forms, templates, policies, procedures, or processes to ensure these practices are repeatable.

Description of Areas for Improvement:

Describe any practices that need improvement, the improvement plan, and any follow-up dates or information for corrective action.

5.6 CONTRACTOR STATUS REPORT

The Contractor Status Report is filled out by the contractor and submitted on a regular basis to the project manager. It tracks status for the current reporting period and provides forecasts for future reporting periods. The report also gathers information on new risks, disputes, and issues. Information can include:

- Scope performance
- Quality performance
- Schedule performance
- Cost performance
- Forecasted performance
- Claims or disputes
- Risks
- Preventive or corrective action
- Issues

This information is generally included in the Project Performance Report compiled by the project manager.

CONTRACTOR STATUS REPORT

Project Title: _____ Date Prepared: _____

Vendor: _____ Contract #: _____

Scope Performance This Reporting Period:

```
[                                                            ]
```

Quality Performance This Reporting Period:

```
[                                                            ]
```

Schedule Performance This Reporting Period:

```
[                                                            ]
```

Cost Performance This Reporting Period:

```
[                                                            ]
```

Forecast Performance for Future Reporting Periods:

```
[                                                            ]
```

Claims or Disputes:

```
[                                                            ]
```

Risks:

```
[                                                            ]
```

CONTRACTOR STATUS REPORT

Planned Corrective or Preventive Action:

Issues:

Comments:

CONTRACTOR STATUS REPORT

Project Title: _____   Date Prepared: _____

Vendor: _____   Contract #: _____

Scope Performance This Reporting Period:

Describe progress on scope made during this reporting period.

Quality Performance This Reporting Period:

Identify any quality or performance variances.

Schedule Performance This Reporting Period:

Describe whether the contract is on schedule. If ahead or behind, identify the cause of the variance.

Cost Performance This Reporting Period:

Describe whether the contract is on budget. If over or under budget, identify the cause of the variance.

Forecast Performance for Future Reporting Periods:

Discuss the estimated delivery date and final cost of the contract. If the contract is a fixed price, do not enter cost forecasts.

Claims or Disputes:

Identify any new or resolved disputes or claims that have occurred during the current reporting period.

Risks:

List any risks. These should also be in the Risk Register.

CONTRACTOR STATUS REPORT

Planned Corrective or Preventive Action:

Identify planned corrective or preventive actions necessary to recover schedule, cost, scope, or quality variances.

Issues:

Identify any new issues that have arisen. These should also be entered in the Issue Log.

Comments:

Add any comments that will add relevance to the report.

5.7 PRODUCT ACCEPTANCE

Product Acceptance can be done periodically throughout the project, as with a component of a system, or for the project as a whole. There are two aspects of product acceptance:

1. Verify that the product is correct
2. Validate that it meets the needs of the customer

The Product Acceptance form can include this information:

* Requirements
* Method of verification
* Method of validation
* Acceptance criteria
* Status of deliverable
* Sign-off

Use the information from your project to tailor the form to best meet your needs.
The Product Acceptance form can receive information from:

* Requirements Documentation
* Requirements Traceability Matrix
* Scope Baseline

It provides information to:

* Change Requests
* Project Close-out Report

PRODUCT ACCEPTANCE FORM

Project Title: _____   Date Prepared: _____

ID	Requirement	Verification Method	Validation Method	Acceptance Criteria	Status	Sign-off

PRODUCT ACCEPTANCE FORM

Project Title: _____   Date Prepared: _____

ID	Requirement	Verification Method	Validation Method	Acceptance Criteria	Status	Sign-off
Identifier.	Describe the requirement.	Method of verifying the requirement is met.	Method of validating the requirement meets the stakeholder's needs.	Criteria for acceptance.	Accepted or not.	Signature of party accepting the product.

Closing

6.1 CLOSING PROCESS GROUP

The purpose of the Closing Process Group is to complete contracts, project work, product work, and project phases in an orderly manner. There are two processes in the Closing Process Group:

- Close Project or Phase
- Close Procurements

The intent of the Closing Process Group is to at least

- Close all contracts
- Close project phases
- Document lessons learned
- Document final project results
- Archive project records

As the final processes in the project, the closing processes ensure an organized and efficient completion of deliverables, phases, and contracts.

The forms used to document project closure include:

- Procurement Audit
- Contract Close-out
- Project Close-out
- Lessons Learned

6.2 PROCUREMENT AUDIT

The Procurement Audit is a structured review of the procurement process. Information in the audit can be used to improve the process and results on the current procurement or on other contracts. Information recorded in the audit includes:

- Vendor performance audit
  - Scope
  - Quality
  - Schedule
  - Cost
  - Other information, such as how easy the vendor was to work with
- Procurement management process audit
  - Process
  - Tools and techniques used
- Description of good practices
- Description of areas for improvement

This information can also be used to collect information for Lessons Learned. The information can be combined with the Contract Close-out report or used separately. Use the information from your project to determine the best approach.

A Procurement Audit is a tool from the process 12.4 Close Procurements in the *PMBOK® Guide*—Fourth Edition.

PROCUREMENT AUDIT

Project Title: _____   Date Prepared: _____

Project Auditor: _____   Audit Date: _____

Vendor Performance Audit

What Worked Well:	
Scope	
Quality	
Schedule	
Cost	
Other	
What Can Be Improved:	
Scope	
Quality	
Schedule	
Cost	
Other	

Procurement Management Process Audit

Process	Followed	Tools and Techniques Used
Plan Procurements		
Conduct Procurements		
Administer Procurements		
Close Procurements		

Description of Good Practices to Share:

Description of Areas for Improvement:

PROCUREMENT AUDIT

Project Title: _____ DatePrepared:_____

Project Auditor: _____ Audit Date: _____

Vendor Performance Audit

What Worked Well:	
Scope	*Describe aspects of product scope that were handled well.*
Quality	*Describe aspects of product quality that were handled well.*
Schedule	*Describe aspects of the project schedule that were handled well.*
Cost	*Describe aspects of the project cost that were handled well.*
Other	*Describe any other aspects of the procurement that were handled well.*
What Can Be Improved:	
Scope	*Describe aspects of the product scope that could be improved.*
Quality	*Describe aspects of the product quality that could be improved.*
Schedule	*Describe aspects of the project schedule that could be improved.*
Cost	*Describe aspects of the project cost that could be improved.*
Other	*Describe any other aspects of the procurement that could be improved.*

Procurement Management Process Audit

Process	Followed	Tools and Techniques Used
Plan Procurements	*yes or no*	*Describe any tools or techniques that were effective for the process.*
Conduct Procurements	*yes or no*	*Describe any tools or techniques that were effective for the process.*
Administer Procurements	*yes or no*	*Describe any tools or techniques that were effective for the process.*
Close Procurements	*yes or no*	*Describe any tools or techniques that were effective for the process.*

Description of Good Practices to Share:

Describe any good practices that can be shared with other projects or that should be incorporated into organization policies, procedures or processes. Include information on lessons learned.

Description of Areas for Improvement:

Describe any areas that should be improved with the procurement process. Include information that should be incorporated into policies, procedures or processes. Include information on lessons learned.

6.3 CONTRACT CLOSE-OUT

Contract Close-out involves documenting the vendor performance so that the information can used to evaluate the vendor for future work. Additionally, information from the Contractor Status Report can be used when collecting information for Lessons Learned. Before a contract can be closed out, all disputes must be resolved, the product or result must be accepted, and the final payments must be made. Information recorded as part of Contract Close-out includes:

- Vendor performance analysis
  - Scope
  - Quality
  - Schedule
  - Cost
  - Other information, such as how easy the vendor was to work with
- Record of contract changes
  - Change ID
  - Description of change
  - Date approved
- Record of contract disputes
  - Description of dispute
  - Resolution
  - Date resolved

The date of contract completion, who signed off on it, and the date of the final payment are other elements that should be recorded.

The Contract Close-out report can be combined with the Procurement Audit report. This information can be used in the Lessons Learned document and the Project Close-out report. Use the information from your project to determine the best approach.

CONTRACT CLOSE-OUT

Project Title: _____   Date Prepared: _____

Project Manager: _____   Contract Representative: _____

Vendor Performance Analysis

What Worked Well:	
Scope	
Quality	
Schedule	
Cost	
Other	
What Can Be Improved:	
Scope	
Quality	
Schedule	
Cost	
Other	

Record of Contract Changes

Change ID	Change Description	Date Approved

Record of Contract Disputes

Description	Resolution	Date Resolved

Date of Contract Completion _____

Signed Off by _____

Date of Final Payment _____

CONTRACT CLOSE-OUT

Project Title: _____   Date Prepared: _____

Project Manager: _____   Contract Representative: _____

Vendor Performance Analysis

What Worked Well:	
Scope	Describe aspects of product scope that were handled well.
Quality	Describe aspects of product quality that were handled well.
Schedule	Describe aspects of the project schedule that were handled well.
Cost	Describe aspects of the project cost that were handled well.
Other	Describe any other aspects of the procurement that were handled well.

What Can Be Improved:	
Scope	Describe aspects of the product scope that could be improved.
Quality	Describe aspects of the product quality that could be improved.
Schedule	Describe aspects of the project schedule that could be improved.
Cost	Describe aspects of the project cost that could be improved.
Other	Describe any other aspects of the procurement that could be improved.

Record of Contract Changes

Change ID	Change Description	Date Approved
ID	Briefly describe the change. Refer to the change log if necessary.	Date signed off

Record of Contract Disputes

Description	Resolution	Date Resolved
Describe any claims or disputes	Describe the resolution including any arbitration or dispute resolution	Date signed off

Date of Contract Completion _____

Signed Off by _____

Date of Final Payment _____

6.4 PROJECT CLOSE-OUT

Project Close-out involves documenting the final project performance as compared to the project objectives. The objectives from the Project Charter are reviewed and evidence of meeting them is documented. If an objective was not met, or if there is a variance, that is documented as well. In addition, information from the Contract Close-out is documented. Information documented includes:

- Project description
- Project objectives
- Success criteria
- How met
- Variances
- Contract information
- Approvals

The Project Close-out report is related to the Contract Close-out report and the Lessons Learned documentation. This information can be combined for smaller projects. Use the information from your project to determine the best approach.

PROJECT CLOSE-OUT

Project Title: _____

Date Prepared: _____

Project Manager: _____

Project Description:

Project Objectives	Success Criteria	How Met	Variance
Scope:			
Time:			
Cost:			
Quality:			
Other:			

PROJECT CLOSE-OUT

Contract Information:

Approvals:

_____
Project Manager Signature

_____
Project Manager Name

_____
Date

_____
Sponsor or Originator Signature

_____
Sponsor or Originator Name

_____
Date

PROJECT CLOSE-OUT

Date Prepared: _____   **Project Manager:** _____

Project Title: _____

Project Description:

Provide a summary-level description of the project. This information can be copied from the Project Charter.

Project Objectives	Success Criteria	How Met	Variance
Scope:			
A statement that describes the scope needed to achieve the planned benefits of the project.	*The specific and measureable criteria that will determine project success.*	*Provide evidence that the success criteria was met.*	*Explain any scope variances.*
Time:			
A statement that describes the goals for the timely completion of the project.	*The specific dates that must be met to determine schedule success.*	*Identify the date of final delivery. Use the information from the Product Acceptance form.*	*Explain any schedule or duration variances.*
Cost:			
A statement that describes the goals for the project expenditures.	*The specific currency or range of currency that defines budgetary success.*	*Enter the final project costs.*	*Explain any cost variances.*
Quality:			
A statement that describes the quality criteria for the project.	*The specific measurements that must be met for the project and product to be considered a success.*	*Enter the verification and validation information from the Product Acceptance form.*	*Explain any quality variances.*
Other:			
Any other types of objectives appropriate to the project.	*Relevant specific measureable results that define success.*	*Enter the evidence that other objectives have been met.*	*Explain any other variances.*

PROJECT CLOSE-OUT

Contract Information:

Provide information on contracts. Enter information from the Contract Close-out report.

Approvals:

_____
Project Manager Signature

_____
Project Manager Name

_____
Date

_____
Sponsor or Originator Signature

_____
Sponsor or Originator Name

_____
Date

6.5 LESSONS LEARNED

Lessons Learned can be compiled throughout the project or at specific intervals, such as the end of a life cycle phase. The purpose of gathering Lessons Learned is to identify those things that the project team did that worked very well and should be passed along to other project teams and to identify those things that should be improved for future project work. Lessons Learned can be project oriented or product oriented. They should include information on risks, issues, procurements, quality defects, and any areas of poor or outstanding performance. Information that can be documented includes:

- Project performance analysis
  - Requirements
  - Scope
  - Schedule
  - Cost
  - Quality
  - Human resources
  - Communication
  - Stakeholder management
  - Reporting
  - Risk management
  - Procurement management
  - Process improvement
  - Product-specific information
- Information on specific risks
- Quality defects
- Vendor management
- Areas of exceptional performance
- Areas for improvement

This information is used to improve performance on the current project (if done during the project) and future projects. Use the information from your project to tailor the form to best meet your needs.

LESSONS LEARNED

Project Title: _____   Date Prepared: _____

Project Performance Analysis

	What Worked Well	What Can Be Improved
Requirements definition and management		
Scope definition and management		
Schedule development and control		
Cost estimating and control		
Quality planning and control		
Human resource availability, team development, and performance		
Communication management		
Stakeholder management		
Reporting		
Risk management		
Procurement planning and management		
Process improvement information		
Product-specific information		
Other		

LESSONS LEARNED

Risks and Issues

ID	Risk or Issue Description	Response	Comments

Quality Defects

Description	Resolution	Comments

Vendor Management

Vendor	Issue	Resolution	Comments

Other

Areas of Exceptional Performance	Areas for Improvement

LESSONS LEARNED

Project Title: _____   Date Prepared: _____

Project Performance Analysis

	What Worked Well	What Can Be Improved
Requirements definition and management	List any practices or incidents that were effective in defining and managing requirements.	List any practices or incidents that can be improved in defining and managing requirements.
Scope definition and management	List any practices or incidents that were effective in defining and managing scope.	List any practices or incidents that can be improved in defining and managing scope.
Schedule development and control	List any practices or incidents that were effective in developing and controlling the schedule.	List any practices or incidents that can be improved in developing and controlling the schedule.
Cost estimating and control	List any practices or incidents that were effective in developing estimates and controlling costs.	List any practices or incidents that can be improved in developing estimates and controlling costs.
Quality planning and control	List any practices or incidents that were effective in planning, assuring, and controlling quality. Specific defects are addressed elsewhere.	List any practices or incidents that can be improved in planning, assuring, and controlling quality. Specific defects are addressed elsewhere.
Human resource availability, team development, and performance	List any practices or incidents that were effective in working with team members and developing and managing the team.	List any practices or incidents that can be improved in working with team members and developing and managing the team.
Communication management	List any practices or incidents that were effective in planning and distributing information.	List any practices or incidents that can be improved in planning and distributing information.
Stakeholder management	List any practices or incidents that were effective in managing stakeholder expectations.	List any practices or incidents that can be improved in managing stakeholder expectations.
Reporting	List any practices or incidents that were effective in reporting project performance.	List any practices or incidents that can be improved in reporting project performance.
Risk management	List any practices or incidents that were effective in the risk management process. Specific risks are addressed elsewhere.	List any practices or incidents that can be improved in the risk management process. Specific risks are addressed elsewhere.
Procurement planning and management	List any practices or incidents that were effective in planning, conducting, and administering contracts.	List any practices or incidents that can be improved in planning, conducting, and administering contracts.
Process improvement information	List any processes that were developed that should be continued.	List any processes that should be changed or discontinued.
Product-specific information	List any practices or incidents that were effective in delivering the specific product, service, or result.	List any practices or incidents that can be improved in delivering the specific product, service, or result.
Other	List any other practices or incidents that were effective, such as change control, configuration management, etc.	List any other practices or incidents that can be improved, such as change control, configuration management, etc.

LESSONS LEARNED

Risks and Issues

ID	Risk or Issue Description	Response	Comments
Identifier.	Identify specific risks that occurred that should be considered to improve organizational learning.	Describe the response and its effectiveness.	Indicate what should be done to improve future project performance.

Quality Defects

ID	Defect Description	Resolution	Comments
Identifier.	Identify quality defects that should be considered to improve organizational effectiveness.	Describe how the defects were resolved.	Indicate what should be done to improve future project performance.

Vendor Management

Vendor	Issue	Resolution	Comments
List the vendor.	Describe any issues, claims, or disputes that occurred.	Describe the resolution.	

Other

Areas of Exceptional Performance	Areas for Improvement
Identify areas of exceptional performance that can be passed on to other teams.	Identify areas that can be improved on for future projects.

About the CD-ROM

This appendix provides you with information on the contents of the CD that accompanies this book. For the latest and greatest information, please refer to the ReadMe file located at the root of the CD.

SYSTEM REQUIREMENTS

- A computer with a processor running at 120 Mhz or faster
- At least 32 MB of total RAM installed on your computer; for best performance, we recommend at least 64 MB
- A CD-ROM drive

USING THE CD

To access the content from the CD, follow these steps:

1. Insert the CD into your computer's CD-ROM drive. The license agreement appears.

    Note to Windows users: The interface won't launch if you have autorun disabled. In that case, click Start-->Run (For Windows Vista, Start-->All Programs-->Accessories-->Run). In the dialog box that appears, type D:\Start.exe. (Replace D with the proper letter if your CD drive uses a different letter. If you don't know the letter, see how your CD drive is listed under My Computer.) Click OK.

    Note for Mac Users: The CD icon will appear on your desktop, double-click the icon to open the CD and double-click the "Start" icon.
2. Read through the license agreement, and then click the Accept button if you want to use the CD.
3. The CD interface appears. The interface allows you to install the programs and run the demos with just a click of a button (or two).

WHAT'S ON THE CD

This companion CD has an electronic version of every form in this book. All the forms are in Microsoft Office format, and all of them can be tailored. Most forms are in Microsoft Word, although some are in Microsoft Excel. The forms are arranged by process group:

- Initiating: 4 forms
- Planning: 32 forms
- Executing: 10 forms
- Monitoring and Controlling: 6 forms
- Closing: 4 forms

In some instances, such as the responsibility assignment matrix, the sample form is included, but in practice you will have to create your own. The sample is only to show how a form might look.

TROUBLESHOOTING

If you have difficulty installing or using any of the materials on the companion CD, try the following solutions:

- **Turn off any anti-virus software that you may have running.** Installers sometimes mimic virus activity and can make your computer incorrectly believe that it is being infected by a virus. (Be sure to turn the anti-virus software back on later.)
- **Close all running programs.** The more programs you're running, the less memory is available to other programs. Installers also typically update files and programs; if you keep other programs running, installation may not work properly.
- **Reference the ReadMe.** Please refer to the ReadMe file located at the root of the CD-ROM for the latest product information at the time of publication.

CUSTOMER CARE

If you have trouble with the CD-ROM, please call the Wiley Product Technical Support phone number at (800) 762-2974. Outside the United States, call 1 (317) 572-3994. You can also contact Wiley Product Technical Support at **http://support.wiley.com**. John Wiley & Sons will provide technical support only for installation and other general quality control items. For technical support on the applications themselves, consult the program's vendor or author.

To place additional orders or to request information about other Wiley products, please call (877) 762-2974.

Index